# ENERGY, ENVIRONMENT, & ECONOMY

# TABLE OF CONTENTS

# FOREWORD BY THE EDITOR

America, land of the brave and home of the free. The United States - A Federation of states bound by the supreme law of the land - the Constitution. The USA, a grand example of what can be accomplished in less than 3 centuries. The world's sole remaining "Superpower", a bastion of hope and freedom that all the world envies. The United States - the largest provider, by far, of assistance for countries and peoples in need. - When there are natural calamities, anywhere in the world, it's the US that's always there to help.

The world's police force, savior of the downtrodden, and bearer of the lantern of hope for all who cry out for mercy and justice. The US of A. - Once home to the mightiest industries known to man. Home of science and technology and miracle cures. Land of oil gushers and railroads and mighty ships. We went to the moon. We were makers of Steel and cars and electronics and aircraft. This was the one place on earth where a man could wish a wish and have a chance to fulfill his dreams. A place where the sky was the limit and all had the opportunity to excel.

We used to manufacture refrigerators and hand tools, clothes and lamps, shoes, and camping supplies, toasters and drywall, picture frames and outboard engines. We made can openers and car seats.

Have we come to the beginning of the end? Is our best behind us?

Our standard of living is no longer number one on the planet, although it once was. Our jails and prisons are over-crowded, and the percentage of incarcerations per capita continuously climbs to new levels. America is not the richest country on the planet, but it has been in the past. We are being over-run by uninvited criminal trespassers, that some would have us call "undocumented immigrants" (we wouldn't want to hurt their feelings!) Many abroad now hate us for our wealth and alleged arrogance. Terrorists scheme for our immediate downfall.

Our infrastructure is crumbling. Our bridges and highways, our water and sewer lines, and our power grids are old and decaying.

Our government is broke, the citizenry is bankrupt and dis-heartened, and some would have us believe that we, as a nation, have already seen our best days.

We no longer make many things. We borrow money from our international rivals to buy their sub-standard and unsafe products. We have mortgaged the futures of our children, and their children.

Where did it go? What horrible misfortune has caused this apparently imminent demise? It has been neither drought nor plague nor revolution. It has not been the mis-alignment of the stars, nor has it been the result of GOD's damnation. For GOD has truly blessed our country with more resources and diversity of talents, more blessings and freedoms than any other society in history. Our decline has not been through the armed invasion of a foreign power, nor has it been due to a lack of caring.

We don't drill for oil much, anymore. And the railroads are all but gone. Our auto industry needs help from the government, and now the Labor Unions are mega-shareholders. We stopped going to the moon. And worst of all, the hope, the dream of 300 million, the belief that we could do anything, because we were AMERICANS, is fading away. How terribly sad, how heart breaking that this should be happening on my watch. Our watch.

Where did it go? What has changed over the decades since the times when things were better? Since the times when WE were better?

I believe we have lost our way. We have strayed from the things that worked, the things our Founding Fathers risked their lives and fortunes to guarantee. We have lost our vision, our identity as Americans.

There are those, now in power, who believe that we're going in the right direction as a nation, that if we "stay the course", everything will be just fine. I don't believe them. In truth, I believe the exact opposite. I believe that if we are to restore America to its former glory, we MUST return to the values, laws, and mode of government that worked so well in the past.

Madison and Jefferson and Washington and Adams and Henry and Franklin and Paine and all the others - they knew. They were wise men, learned men, men who read and men who did their homework. They were collectors of books and were avid debaters. They forged a vision of a governing body which would enable the states and their citizens to go about their business without interference from Kings or dictators or other power brokers. In America, they thought, we can all be kings. We can all chart our own destiny with the guarantee that no government; no authority could revoke or infringe upon our God-given rights.

They formed a federation of states, a Federal government, as a cooperative venture between the states, since there were a few areas of governance which individual states could not administer alone. Defense of the country and common currency were perhaps the top two on their list. The states jointly ratified a "Constitution", which enumerated the rights of its citizens, and thereby assured that all the inhabitants of all the states had universal protection from interference by this new Federation. This Constitution (and its original appendix - the Bill of Rights), stated, more than once, that this was the sum total of the deal - Individual freedom would prevail, the states were the prime form of governance, and the roles of the new federation were clearly enumerated, limited, and there primarily to protect the rights of individuals. Carbon Credits, salary caps, "Obama phones", banking cartels (a.k.a. the Federal Reserve), fishing licenses, gun registration, Federal databases of its citizens, and portion control of soft drinks were not part of their list. Actually, these things were on their list as activities banned by the new Federal Government, since they specifically stated that Federal powers not listed in the Constitution were banned.

I have NO DOUBT that our Founders would be appalled at the monster their child has become.

I believe that one of the major failings of our current Republic is the continual, persistent accumulation of power by the Federal Government. This power has been wrested from the states, the localities, and the individuals of this country. Further, it would appear that even within the structure of the Federal system, there has been additional power grabbing occurring in recent decades. The Executive Branch has apparently chosen to ignore the "balance of power" concept fundamental to our system, and taken it upon itself to decide which laws to enforce, who to prosecute (and not prosecute) and further decided that the writing of laws by Congress is unnecessary since the President can do whatever the hell he pleases via "Executive Order".

The difference between this modus operandi and that of a King is far smaller than our Founders would have accepted. Indeed, it is far smaller than we should accept.

Our Founders envisioned Political service as a duty. They thought it a responsibility that most would be loath to accept. They envisioned citizen politicians who would serve as representatives of their community for little pay or glory. And they fully expected that after one term of service these citizens would be anxious to return to their homes and jobs. They expected that men of honor and men of wisdom would fulfill this role out of a sense of duty and love of country.

Why have we allowed this change in the spirit of public service? For generations we have, for the most part, trusted our government and our media. We believed Cronkite and Murrow and Leher. Today, unfortunately, it is becoming more evident that the moguls of media and the career politicians now in power have violated this trust. There should be little doubt that they, if allowed, shall continue to do so. We are being manipulated and deceived, and the truth of this is slowly rising to the consciousness of the national psyche.

I am confident that the predecessors in these fields, at times in the past, also misled and deceived, but I also believe that the ends sought through these deceptions have changed. What once was done in the best interest of the country is now done for other reasons - self-serving reasons. Reasons of power and of money and of glory and of sexual favor. Motives of re-designing our country to their own philosophies.

This, then, is the first area of discussion in this book. - Governmental power, control, taxation, monetary policy, personal and governmental responsibilities, and the importance of Industry.

The second section of this book deals with a major tool mis-used by our current "Leaders" to achieve their ends: The calamity of Climate Change and the environment.

This is not the first such manufactured calamity, nor will it be the last.

This second section examines the "science" now being used as proof of a current calamity. An alleged calamity, which should cause us all to modify our ways of life to fulfill the desires of those who would have more money, more power, more fame, and more sexual favors. And, having learned to trust our government and our media, we of course diligently grant their wishes.

When my long-time friend, Harold, asked me to edit this book, I had no idea of what I was getting into. I knew very little, at that time, about the details of Global Change or $CO_2$ concentrations. Since then I have spent 100's of hours searching the WEB and looking at the data. I have been surprised by the quantity of information available, and to some degree, I have been surprised by the information not available. The "science" being reported by the power-grabbers "proves":

The disastrous impact of "Global Warming" and/or "Global Cooling" that has been caused by mankind will lead to the destruction of life on this planet. This is evidenced by melting ice caps, $CO_2$ rise, ozone hole creation, and sea level rise.

The second half of the book looks at the science behind these claims.

It contains a wealth of information concerning historical and current data relating to $CO_2$, global temperatures and the ozone hole. These chapters contain discussions regarding these topics, and provide information concerning numerous factors, aside from human activity, which lead to global changes.

I hope that you read this book. I hope that when the next "calamity" occurs, you will remember the discussions herein. When we are told that the aliens are coming, or that water causes cancer, or that working 10 hours a day is healthier, (and therefore required of those receiving Obamacare), I hope you will interrupt your busy life long enough to think about what this book says.

If we wish to have any hope of our country remaining free, and of regaining the freedoms we have already lost, we must all make the commitment to stay informed, to use our own powers of reasoning, and to resist those who would manipulate us for their own benefit.

Howard Huegel July, 2013

I approve this message. - Harold Seelig, author, July 2013

# INTRODUCTION

This book is written by a self-professed non-expert, in which we examine aspects about the current state of affairs in our nation, in order to provide a stimulus for rational thought about where we should be heading. These affairs include the economy, the environment, taxes, monetary policy, industry, social responsibility, government and politics, and this book contains discussions pertaining to these topics.

Any book invariably includes not only the thoughts and ideas of the author, but additionally cultural norms and the spirits and thoughts of countless others, whether directly contributing to the work or not.

This work is intended to stimulate a desire to understand these and other subjects and to prompt the reader to make their own rational decisions about them. This work emphasizes the importance of "The Will of the American People", and the significance of this critical force in maintaining Freedoms. The American People and American Businesses are tired of being deceived, manipulated and bullied by their government.

The message of this book is that each of us has a mind, and we need to use our minds to make this country a better place. Those who give up their right to think for themselves wind up being someone else's lackey, a virtual slave.

We are an advanced society, having graduated over countless generations, from cavemen to tribes and villages, city-states, and fiefdoms, through monarchies and dictatorships to polite and civil societies. As we've graduated to each higher level of refinement and each improvement in the condition of life, there are always some who are left behind, some unwilling to take responsibility for themselves, and particularly in taking responsibility for thinking for themselves. Our brains are there to use, to think things through, not just to keep our heads from collapsing. This book examines the chant of 'Human $CO_2$ related Climate Change' and the real reasons for its popularity, which are: 1.) Fear 2.) Hate and 3.) Misguided charity. This conviction of ourselves, or of our industries and our transportation is not unlike the damnification of witches during times of social mental disorders of centuries now past. The commonality between today's perceived crisis and of those in the past are the disorders of unjustified fear and hate, and today also include misguided charity. The science, or rather, lack of science supported by actual data, is herein highlighted. We look at climate change during the times 'before man' and the ways humans can possibly have an effect on Global Climate. We can prove 'Human $CO_2$' is having virtually no effect on global temperature, and therefore no effect on Global Climate. Dr. Paul Joseph Goebbels was right. As Reich Minister of Public Enlightenment and Propaganda, Goebbels revealed that if a lie is told often enough, it will become the new knowledge and will be accepted as truth. This mirrors the real truth of our $CO_2$ 'crisis'. The $CO_2$ 'crisis' is not real science, but rather an emotional crisis stimulated by fear, hate and mis-guided charity.

# SECTION 1 - THE LIVES WE LEAD

## THE RATIONALE OF RATIONALITY

Rational minds are moved when a similar conclusion is reached by a preponderance of unbiased experts. Emotional decisions are often moved by a singular event (a photo of a starving kid with a fly in her eye) which is used as "proof" of some point when presented by a skilled orator - often a popular personality. Sound decisions are rather based on rational thought. Emotions are manipulated as a last ditch effort to rally people to a cause when logic and reason don't or can't work, or have already failed to win support.

We must closely guard against manipulation. More often than not, these efforts are based on emotions of emergency. Of course humans and all animals benefit from emotions of emergency. For instance, an alarm call such as "lion" ignites the "fight or flight" adrenaline rush for self-preservation. Likewise "tsunami" alerts people to get away from a shoreline. In such emergencies, it is imperative for people to immediately respond without thinking. Today, the proponents of gloom and doom are screaming "DANGER", trying to convince us we are on the cusp of triggering an irreversible breakdown. (For instance, a breakdown of the climate, and that we may NOT take time, due to imminent calamity, to listen to the voices of real unbiased science and reason).

If we look objectively at the progress of a society without getting caught up in the deliberately emotional frenzy of "the new wave" of something to fix "the crisis", we can easily see how reason and control are usurped, always with some sort of rationalization, no matter how tenuous the link with reality - "to save the children", "for the good of the country", "for the good of our religion", "for the good of the society/economy", "for the good of the world" (really ramping up) and on and on.

Manipulation of hate may well be the strongest coercive tool, convincing one's followers that another society, group, or individual is against theirs, and is in some way responsible for their own failings. This can lead to "blind hate" which blocks out logical thought. Many may recall a recent battle cry of "Voting is the best revenge." - B.H. Obama 11/2/2012. We must ask ourselves 'Revenge for what?' Our anemic economy over the last 4 years? Slavery? The Free Market? American freedoms and a Constitutional Republic? America being the most anti-colonial country in the history of the Earth?

Examples of misdirection of logic in our literature and media abound. "Common Sense" gets us beyond these false "rationale" and of the coercion of emotions by those who would attempt to misdirect us. "The voices of reason" will gradually follow emotional responses to an issue, as logical, reasonable consideration takes some time for people to ponder an issue or event. Rational thought is the foundation of our civilization as compared to the immediacy of decisions based on emotions, as in the world of animals.

MEDIA

We are hearing a lot these days about how our NEWS media seems to favor "The Democratic Party" in commentary and in coverage of the news. We first need to look at the motivation for a media organization. They are structured to be "for profit" groups, operating for the benefit of the owners. A larger audience gives higher ratings, which commands more advertising dollars. So, the objective is to get a larger audience. Items of immediate, large-scale impact on society, emotionally driven acts of violence, fear and tragedy, all command large audiences. Items of great excitement and patterned social disharmony also create high interest.

When we compare the general characteristics of "Liberal" vs. "Conservative" parties, the Liberal is the one more driven for inter-personal and social relationships, more geared toward blame and condemnation, and is more adamant about the importance of human emotions. The Conservative Party is much more focused on the mundane, the repetitive, marching in a logical forward direction for improvement in the standard of living for all. The Conservative view is much more predictable and boring, and is in fact more altruistic by nature. The Conservative view is that we deserve to enjoy the results of our efforts, and we have the right, but not the responsibility, to give personal charity.

So, who gets the biggest, most emotionally charged audience? The Liberal party, of course. Media hang on to the Democrats, the "active little cherubs", and forgive almost everything they do. Media seem to like to condemn the really Rich, except of course those of Liberal bent. When the Democrats rule our House of Representatives, the news is more exciting. So media preference is for Democrats. Recognizing this trend explains a lot of why we see what we see, but this recognition in no way condones this behavior. My belief is that public recognition is it to be expressed for those who make the most stuff, and therefore benefit society the most, that is, the Conservatives. Conservative people tend to be the most admiring of personal achievement, excellence, good work ethic, family centered life, rewards based on contribution, and are in actuality, the least 'racist' of the two major groups.

COERCION

Crucial to understanding "The Economy", "Politics", and other key issues is knowing the underlying strategies of liberal groups. Simply stated, "They" would have us engage our minds with the immediate proclamations, "sound bytes", and coercive or "emergency" remarks of the day - rather than looking at the long term trends of a group, the slips of the tongue, and remarks by peripheral agents. We must look at the effects of particularly big shifts in societal goals (the "emergency" of free health care for all, for instance), and what these will really mean in terms of human incentive, total costs, Productivity, and therefore the health and stability of our whole society.

Attempts to convince people that their condition, their relative wealth, and their comforts in life are inadequate and are someone else's fault should be seen as mere social propaganda, (as are attempts to "stir up stuff" to create bad feelings). Thankfully, these attempts are usually limited to an extremely small number of loud people and their paid minions. The massive numbers of people who use their intelligence in support of Productivity and the enjoyment of the fruits of one's own labors eclipse these splinter groups and their radicals. Unlike those supporting 'micro loans' to start up productive 'cottage industries', a Community Organizer often acts in the spirit of revenge and social disharmony in order to grab money and power.

Being led to believe things to be true that are not true is a total abomination of GOD's Plan. All too many people are involved in generating the "un-true". The extreme sadness of it all is that those people involved in "bending" the truth actually believe they are serving a higher ultimate goal, through a zealous blind faith in their leader(s). "The Ends Justify the Means" and "Workers of the World Unite" are hallmark phrases of these distortions. "We must re-write history" to paint a distortion of the past is as well a bad thing. Book burnings are a similar way of destroying past achievements, knowledge, and a sense of morality and social norms. The Nazi party was a prime promoter of these tactics.

These same devious deceivers often incite to cause fantastic emotional distress and then offer relief from anxiety and pain with their (old) "new" program. Logic and rational thought, however, quickly dispel such fantasies. Trying to turn a long-term

14

socialistic plan into an emergency of epic proportions is but one example. If National Health Care will, for instance, save so many trillions of dollars, it can just pay for itself, without up-front investment, without more taxes, without ravaging society through various "ability to pay" schemes.

There are innumerable strategies employed to emotionally circumvent a person's logic. After all, emotions happen in an instant, and although an important trait for animals' and for humans' early survival, humankind and society have developed due to rational thought. Some other species such as elephants, dolphins, whales, etc. have the ability to consider concepts and make rational decisions, but informed, logical thought is considered core to humanity. We don't just sit with bated breath, waiting for a shaman or other leader to tell us what to think or how to act, we must proceed reasonably, or else succumb to another's agenda.

So, manipulation of Public Opinion through emotions is a very powerful tool to "get your way", whether the "ends" (In the US, condemnation of the rich and the taking of their money) justifies the "means" (manipulation of Public Opinion through less than honorable means), or not. The morality of this manipulation is up to each of us to decide. It is generally reasoned that manipulation is used as a desperate last-ditch effort, after rational, scientific, and honest, logical methods have all failed.

BAD SCIENCE

We'd be negligent to not discuss the use of "Science" when it is used to incite emotions regarding various "policies". This misuse of "Science" is becoming more prevalent in regard to $CO_2$ and is being used to question whether groups and industries re-liberating $CO_2$ should be penalized.

When we look at the real necessities for life, there are actually only 4: Oxygen, Water, Sunlight, and Carbon Dioxide. Without any one of these 4 items, there could not be life on our Earth, the Blue Planet.

In our media (and entertainment), the opportunities to misguide the public are everywhere. In the case of mercury in fish, we have access to results of millions of tests over the past 50 years. The results have been findings of levels of about .7 ppm in swordfish, tuna, etc. This is now being promoted to be the result of coal combustion (which releases a tiny, but now measurable trace amount of mercury). There are, however, some "museum samples" of fish collected prior to the industrial revolution, back in the 19th century, (which do not have mercury contamination in the preservation liquid), which have been tested for mercury, and are all found to have about the same level as recently caught fish. BUT, we have such a preponderance of recent data, and only a few old samples, that the importance of the museum samples is overwhelmed. If we have 3 million samples of contemporary fish, and only 40 museum samples, the relevance of the few samples compared to the many will be questioned, but must be fairly compared in a before-after analysis.

An exciting time recently passed, when a few doomsayers got many people upset and panicky, over the "end" of the Mayan Calendar on December 21, 2012. Of course those doomsayers were not thinking about the wisdom of the Mayans. To keep from wasting time in repeating work they'd already done, I'm sure everyone in Mayan society realized that all they would need to do is rollover to calendar "day one". We don't fret when a clock approaches midnight, do we? We just "automatically" know that we can use the same clock again, for the beginning of the next day. The Mayan's calendar covers short term, long term, and very long term time periods. There is no warning of doom in these calendars. The world did not, in fact, end on that date.

Over millions of years, the planets in our Solar System have all lined up the same way many, many times. It's just doing it again. It's merely time for a big New Year's celebration. The World didn't end the last time this happened; it was not the time of the collection of all good souls and eternal damnation of the remainder. We also did not all vaporize on the rollover of 'Y2K'.


## CONTROL OF INFORMATION

Public Opinion is swayed by many things, including its exposure to truly scientific information. The problem is that in general the public sees only the "scientific proof" that someone wants the public to see, perhaps for that same someone's own personal interests. For instance, North Korea has a tremendous control over what North Koreans are allowed to experience, particularly in terms of information from "other lands". With no other information to learn and choose from, they can only believe what they hear, whether it's right or wrong. Additionally, there are probably not too many people willing to freely disagree and share discordant ideas with others in North Korea.

Another example; during the "cold war", scientists, even those of significant position in The Soviet Union, were extensively trained to suspect their peers in other countries (particularly in the USA), and most particularly when it related to the general standard of life in the US.

My father was involved as an "International Cooperation Officer" between the US and a number of other countries regarding efforts for scientific research in the Antarctic, (and, incidentally, has two mountains in the Antarctic named in his honor). While living in the Washington, DC area, he and my Mother often entertained visiting foreign scientists in our home. One particular visit in the 1970's was by a Russian Polar research scientist, who wished to visit the Washington, DC area. My parents took him to a number of National Landmarks, museums, and so on in the Washington, DC area. In particular, the Russian Scientist was interested in "riding through" typical American neighborhoods, which of course they did. During the course of discussion one evening, my father explained to the visitor that each of us three children had good jobs and each owned their own homes and each had several cars. The visitor's attitude was very much skeptical, confirming his homeland's training which convinced him Americans would be very convincing in their lies about how good the standard of living in the US actually was. (My father wisely did not tell this scientist that I had my own private plane.) At any rate, the scientist continued his visit to several other US destinations, and returned to visit my parents after realizing that Americans were not, and did not have any reason to try to be deceitful toward him. He apologetically left the US, having learned much more than was intended. The most powerful message he carried home is that his own leaders wanted to deceive him about Americans.

How many of our global neighbors have similar propaganda forced upon their citizens? Why? To hide their own inadequacies? To try to condemn a capitalistic Judeo-Christian nation?

YOU'RE TOO DUMB TO UNDERSTAND

I cannot imagine how many "new wave", "agents of change", and "sweeping revolutionaries" have told "the people" that a situation or a concept or aspect of a society is "too complicated for them to understand". Oh my! Oh, it is so difficult to understand... Surely the first instance of use of terms similar to this had to be when one cave man asked another where the best fishing spot was. It's such an easy and deceptive manipulation of one person or group of people against another.

Why would a person say this? What is their real meaning? We can guess about the various real thoughts of the speaker:

a) "We ourselves don't know" (and don't want to create panic) <probably OK>

b) "We don't know and don't want you to know either" <ego>

c) "We know and don't want you to know". <perhaps evil>

d) "We know and you are too stupid to understand". <demeaning>

e) "We don't know and you are too stupid to understand". <undeserving>

f) "I'm really in a hurry and have better things to do than explain this to you." <I am more important than you - ego>

Hearing an invocation of the phrase, "it's too complicated for you to understand", should create an immediate objective sense of inquiry, it should "ring a warning bell" in the mind of the listener.

The most troubling and uninformed or deceitful application of this coercion must be in the announcement of a plausible Global Government, a.k.a. "a New World Order". This has always signaled the start of destruction of an existing society; to trample over the wishes and wills of the majority; to advance a new group; to be in charge, and to take power/money/control from others. This is often done just prior to silencing the voices of the artists and authors, philosophers and even rational technical specialists. These are time of book burnings, and rewritings of history, as we saw in Nazi Germany, and Mao's Cultural Revolution.

Those, who through proclamations or intensive "teachings", brutal brainwashing, and denials of our right to think in our own way and attempt to force us to accept ideologies without question, are bad people.

I'm not about to just go and do whatever it is that someone has told me to do. I'm going to look at it first, before jumping in. It all boils down to whether it is something "They" want, or something "I" want, doesn't it? Of course "They" want me to believe that "I" want what "They" want, now don't they? Even when what I want is for the success of our economy and a good living for the majority of Americans, "They" say that this can only occur through massive taxation. In Economics 101 we are in fact taught that increasing taxes chokes the market driven recirculation of money, which powers a free economy. Government CANNOT force recirculation of money in a Free Market. Every time this is tried, the very spirit of the Free Market dies.

## INDIVIDUAL RESPONSIBILITY

"Our Constitution was made only for a moral and religious people. It is wholly inadequate to the government of any other." - John Adams

What is each person's responsibility to their society? Let's look first at some rules for living, attributed to GOD and various deep thinkers across the millennia. The Ten Commandments are a good start. Here is the text on the granite monument "Presented to the people and youth of Texas by the Fraternal Order of Eagles of Texas 1961"

"I AM the LORD thy GOD.

Thou shalt have no other GODs before me.

Thou shalt not make to thyself any graven images.

Thou shalt not take the Name of the Lord thy GOD in vain.

Remember the Sabbath day, to keep it holy.

Honor thy father and thy mother that thy days may be long upon the land which the LORD thy GOD giveth thee.

Thou shalt not kill.

Thou shalt not commit adultery.

Thou shalt not steal.

Thou shalt not bear false witness against thy neighbor.

Thou shalt not covet thy neighbor's house.

Thou shalt not covet thy neighbor's wife, nor his manservant, nor his maidservant nor his cattle nor anything that is thy neighbors."


These Ten Commandments are said to be given directly by GOD to Moses, ca. 1,500 BC. These heavenly rules are directed not only to individuals, but also to groups of individuals, and to the governments of those individuals. Similar rules are imbedded in the social character of countless societies; many now long since gone. In each of the Ten Commandments, we see the recognition that common sense and reason, the thought processes and actions that differentiate humans as a special part of the animal kingdom, are all important to improvement of the human condition. These rules particularly decry the actions resultant of wayward emotions. (Greed, envy, lust, etc.)

Literally, these "Ten Commandments" command respect for GOD and GOD's Name, His Day, Our Parents, and Our Individual Rights and Possessions. Particularly, it commands that we must not "Covet", or yearn for another person's possessions. Admiration of achievement of others and for the fruits of that achievement is not greed, it is not envy, but is rather inspirational and leads others to exceptional careers and Productive contributions to Society.

Consider *Mahatma Gandhi*, born Mohandas Karamchand Gandhi, whose pure vision and altruistic concern for Society inspired many hundreds of millions to see that a dominating government and repressive taxation can always be overcome by the Will of The People. The Freedom of body and mind and striving for excellence realized by The People has led India to become highly Productive, improving the material wealth and condition of all. They have taken a deliberate path for promoting the opportunity of Education, respect and creativity, escaping from a virtually crushed Economy and Spirit.

Following are "The Seven Deadly Sins", not offered in a single entry in The Bible. These are:

Wrath

Greed

Sloth

Pride

Lust

Envy

Gluttony


These "Seven Deadly Sins", are frequently being used today as "tools of government" or "tools of politics". Which are purely emotional tools? Of course all are emotions, at their worst unchecked by logic and rational thought. Let's look at these in more detail:

Wrath - Pure, blind anger. Anger for some recent or possibly unknown specific event in the distant past. Lawyers feed from this human malady. Without irrational anger, much more sensible actions and a better society results. The rational response to Wrath is to first use that wondrous capability GOD gave us for thoughtfulness and reason. Then, we plan for a common sense resolution. Of course if someone is at that time harming other people, adrenaline and control are necessities, but not absolute, raging Anger. Anger should be tempered and drained away with common sense.

Greed - The belief that one deserves something they did not earn. A "right" to a life of leisure without having to work, the building of an empire on the backs of un-willing workers, this is Greed. Making a Billion with consenting workers is not Greed. Cheating people out of their money through criminal activity is Greed. Thinking and working hard to get rich is NOT greed. (This is ambition to improve one's situation, not at the cost of others.) To take earnings from those who have worked extra hard to make "extra" money is to discriminate against those of greater Productivity, and this IS greed. Incentive is responsible for the drive to Productivity. Spoken by those not truly understanding this, it is described as "greed". But greed is a dark emotion focused on taking, rather than on making. Most Taxes are a form of government greed. Another form of Greed is the self-righteous compression of power among an elite class.

Sloth - The lack of willingness to take an opportunity, wanting not to work, but to sit back on one's rear end, and particularly demand that others provide support is Sloth. A person deciding on a life of philosophical pursuits or the life of a hermit, while still providing for one's self, is not Sloth.

Pride - A Willingness to take credit or boast beyond a reputation which has been earned. Pride is self-centered and un-earned, an attempt to put oneself above others without an achievement that deserves respect.

Lust - This is simply a natural urge. Human brains are designed to logically put this in perspective. Animals succumb. Animals are not reading this book. Lust is a compelling emotion from the basal brain, necessary for continuation of a species. Simple human morality and thought overcomes this handicap, replacing it with respect, Love, and commitment.

Envy - A distortion of ambition, and closely linked to greed. Envy is an emotional and childish sense that another person has something that that other person does not deserve. This emotion is apparently behind many of the actions of "Liberals". Liberals being those who believe everyone deserves all of the same goods, no matter what the Productivity of each person.

Gluttony - Also closely linked with Greed, Gluttony is consuming beyond that which is necessary, and particularly beyond that which is deserved.

In the Holy Bible, we do not hear that GOD's creatures are intended to sit idly, waiting for food, shelter, etc. to be delivered. Even the "lowly" coral gets up, goes out, and snags a bite to eat. In a hive of social animals, for instance bees, it's true the workers go out and get sweet nectar and bring it back for the others. The others are involved in hive duty, making use of the nectar, pollen, and other materials the workers bring home. And yes, the drones don't have much to do except fly up, and the "best" one gets to mate with the queen, but we all know what happens to all the drones after the queen has been fertilized. They don't lie around waiting to be fed; they're all killed. Only one male and female have the job of making the next generation, and it is literally a hive of one family.

Now if I were a drone, I'd get the heck out of the hive, and go find my own nectar without worrying about siring the next generation.

In a group of animals the adults go out and get dinner for both themselves and their babies, and help to protect the group. Nobody gets to stay around, unless they are contributing, being "Productive" in a substantial way. The exceptions here are those who have already contributed their life's work, and are allowed to enjoy the benefits of retirement.

Of course it is a parent's right to share what they have earned with their families, and the children's right to share what they have earned with their parents and with other family members. It's even OK to share what they have earned with others, as charity and gifts of their own decision. It is a parent's responsibility, accepted at the time of the decision to conceive, to teach their children "the difference between right and wrong", and the responsibilities their children will have in their adult lives. It is each parent's responsibility also, again accepted at the time of the decision to conceive, for providing the physical needs of the child. That responsibility is to provide food, shelter, medical care, education, and a loving environment.

All too often, the messages about right and wrong aren't reaching our children. As a ridiculous example, consider the little kid peaking around the corner of the stairway at the local pimp. He sees great wheels; a big roll of large bills, girls, good times, and the respect of all the other little kids. So that little kid is really motivated to study hard and do well in school to become the next Medical Researcher who may find a cure for cancer, right? Wrong. We must be careful about who we make our Heroes, now shouldn't we? Our true Heroes are the Captains of our

Industries. Just as with any position of power, a person can be led to "the dark side of the force", the side of corruption, the side shifting control of our Country away from our Constitution, away from the timeless superior intellect of our "Founding Fathers". Today, unfortunately, the Captains of our Industries and Private Institutions are being denigrated, and our political "leaders" show little propensity for heroism.

Historically and biblically speaking, charity is rightfully shared with someone lacking food or shelter, or with a pressing need resulting from environmental calamity or medical misfortune, or even a loss from the violent acts of others.

Since money and goods bought represent a part of a working person's life, a charitable person is knowingly and voluntarily giving away part of their life for the benefit of another.

We must not say that it is wrong for any of us to give charity to someone who has made decisions of poor judgment. "The Salvation Army", one of our finest charitable organizations, works very hard to help those who are "down and out", those who have made poor decisions in their lives, and this charitable organization helps them regain their confidence and Productivity.

We must distinguish the various reasons people are "in need", before applying the label of "beggardom". Some people have been very productive, and they need a short bridge to new productive employment. Some have been productive and develop medical trouble, or have become too old to work. Of course none of us want to see poverty and starvation. It is up to each of us to plan for our future. It's the core of the story of 'The Ant and the Grasshopper'.

SOME EXAMPLES

Did Bill Gates merely manipulate a huge inheritance? Did he emotionally manipulate or steal from numerous rich "targets"? Was he the recipient of questionable government funds? No, no, and no. Bill Gates and his associates provided THE WORLD with access to the wonders of the information age. True, he acted at the right time in history. Bill had a vision. He followed that vision. The worth of the work Bill and his associates gave to the world is many, many, many times what they have been paid. What is Bill doing with his honestly earned Billions? He's helping the world. He's fighting Malaria and other global problems. This is true charity, and is the model of what each of us should secretly or openly aspire to: To earn more than enough for our family and ourselves, and to reach out to help others. To contribute to individuals, to "non-profit" charitable organizations, and to other groups who are honestly pursuing that which many of us want to support. This is our job as individuals, not the government's job - to give charity to support individuals, and organizations such as ACORN®, Family Planning®, the SPCA®, the Arts, the Catholic Church, etc.

Should we condemn the honest rich? Should we create a perception that the virtues and benefits of capitalism and a free market are a corruption of the human mind and moral ideals? Should we claim the honest rich have exploited the poor and made them that way? Have Oprah Winfrey or Bill Gates oppressed certain groups of people, the poor or the "disadvantaged", or have they instead created opportunities for Productivity and given ample charity?

Perhaps Bill and Oprah and other highly successful people want to give charity to others. This is what the American system is about. But, do I want a rich person or anyone else telling me who I must give charity to? I don't mind a suggestion, but I do mind being emotionally manipulated or forced. I do mind discriminatory taxes being taken for public charity, or "welfare". Charity is rather the job of the Church, non-profit groups and of personal considerations.

The overwhelming message here is that people need to work for their keep, to earn the right to be proud of their achievements, and to enjoy life and to pursue happiness. We have many sayings for this, such as "pulling one's own weight", "earning one's own keep", and being "responsible for one's own affairs".

We read that Jesus took two fish and five loaves of bread, and fed thousands. This was a miracle and proof of the Glory and Mystery of the power of GOD. We also read that if you give a person a fish, you only feed them for a day, but if you teach a person to fish, then they can feed themselves for their lifetime. A person feeding himself or herself is good. The Holy Bible tells us that everyone being productive is moral and righteous. It is moral and righteous for those who are more productive to give of their own charity, not to be forced by demand or duress.

So, what is each person's responsibility and what does each person deserve? Each person is responsible for providing for themselves and their family and to be Productive and particularly not be a Burden on Society and/or the government. The teachings of The Holy Bible tell us that we each need to be charitable, to provide help on an individual basis to those who have had "bad luck". "Luck" being differentiated from "poor decisions" and "poor judgment". But it is acceptable to offer personal charity to those whose own decisions put them in their situation. The responsibility for Charity belongs to each person, not to society or Government.

A Government should never be in a position to make decisions for individuals about their personal charity. It is up to each one of us to make our own decisions about how much, and to whom we give our charity. It is not up to a Government to take our money to give to someone else. This "power", throughout history, is inexorably linked to corruption and subjugation.

# INSTITUTIONAL AND GOVERNMENTAL RESPONSIBILITY

### GOVERNMENTAL CHARITY

Politics are based upon individual and group beliefs. Some believe everyone deserves the same goods as all others, regardless of their contribution to the Economy and Society. Various religious documents tell us we should give charity of our own will to the poor. None of these documents suggest that people should plan and depend on charity as their life support. Dependence upon charity should always be viewed as a sign of failure; the lack of ability to support one's self or one's family, for whatever reason.

### PRODUCTIVITY ENABLES CHARITY

Always remember the more stuff that is made in a country, the more stuff there is for everyone, and the greater the opportunities there are for everyone, regardless of the "money system". And yes, some people will be more Productive, and will therefore have more and better goods and more "money". One farmer may take more time preparing his/her land than others, grow "green manure" in off seasons, use the right seeds and grow the right crops, and therefore have more food to share with his family and others, however he or she decides. Highly Productive people have an exceptional history of meaningful charity, particularly those in the U.S.

### INCENTIVE/REWARD

Incentive is the promise of a "reward" or payment for doing something. If you hold up a treat in front of a dog, in reality making a social contract with the dog to perform in some certain way, and the dog performs, then the dog gets the treat. If the dog performs and gets the treat, and later performs for another treat, and the treats are shown, but not given, the dog will quickly see the change in the rules, a breaking of the social contract, and not perform with enthusiasm or at all, (or maybe it will get really pissed off). A mule will put in a good day's work for a farmer until the farmer starts cutting off the food. Bees will continue to return to the same flowers to get more nectar, until there is no more nectar. This is incentive and reward. Will the dog want to do some trick to please an owner if the treats are available if the dog performs or not? Will the mule work if it can otherwise get all it wants to eat? Give them all they want, and they don't have to do crap.

Individual people are no different.

Just as we are allowed the opportunity to excel, so are we allowed the opportunity to fail. Knowing of the possibility of failure is one of the many necessary incentives in our understanding of the necessity to Produce. We see many examples of people who perform, and of many who fail. A person who sincerely thinks that we can force a society in such a way that all can succeed has morphed their thinking into a special kind of destructive mental illness. Everyone being Productive and working hard only happens in the "Star Trek®" series, which is fictional.

Incentive is the motivation to do something. If some people need their hair cut, and are willing to pay a competitive price for it, then someone will be willing to cut hair. If people get hungry and want something to eat, then someone will be willing to supply food to them at a competitive price. If only one person is selling food, they can increase their prices and make more money. "Free Enterprise" is the response to the Incentive for someone who is not selling food to start doing so. This competition lowers the price. If too many start selling food, then the prices continue to go down, until only the most efficient can continue to sell food without losing money. This is the natural Free Enterprise System, and it's based purely on Incentive.

If a government begins controlling incentives and rewards by oppressive regulations and taxation/subsidies, or by artificial control of resources or labor, then the Incentives and Rewards are not natural, and the results and spirit of this enterprise are subject to system breakdown and corruption. This can be seen as the root cause of our current economic malaise. The Federal Government is, after all, a monopoly; while State Governments are not. State Governments compete for residents and businesses.

GOVERNMENT LIMITED INCOME AS A NEGATIVE INCENTIVE?

Today we see laws and party politics that go against "The Ten Commandments". A person's income is part of that person's possessions. How can we have laws and "Pay Czars", which would limit income? If a person makes a contract, and is promised a certain income for certain work, how can that person not be paid that amount of money, unless through breaking one of the Ten Commandments?

In our media circus and in our Government, individuals are

being denigrated, and their work contracts voided.

## GOVERNMENT DEINCENTIVIZATION THROUGH CHARITY (WELFARE)

"Charity is no part of the legislative duty of the government."
- James Madison

The forcing by ACORN® and others to give loans to high risk borrowers, who couldn't even repay "good" loan terms, for some sort of entitlement or socially redeeming purpose, is totally against good common sense. And now those who bent to the political pressure are vilified and stripped of their honest income, and their Corporations have been pillaged as though by Pirates. This is most certainly a case of coveting another's money. This is at best, some sort of rationalization for "Economic Equality". "Economic Equality", now promoted as a type of unalienable right, is really a cryptic distortion of "All men are created equal". People are created equally, not entitled to the same goods and living standards.

This distortion of our free enterprise system was due to an emotional, not rational, feeling of pity for those individuals who failed to learn, who failed to gain the sense of a work ethic, who instead learned other talents, or succumbed to various agents of immediate gratification. These individuals gave in to very short-term goals, rather than having the vision and guidance to see the benefit of the long-term goals. Why work hard and save for the future, when there's plenty of distractions and pleasures for today and tonight? Why wait to get married and have children when our system rewards having children out of wedlock? With "Section 8" housing, food stamps, free cell-phones, fuel assistance, and a multitude of other "don't work" incentives, why should committing oneself to the drudgery of a life of work even be considered? And what is the parental model presented to these people's children? Duhhhhhh.

Our welfare "public charity" rolls are expanding at an alarming rate, and our Welfare system has made it so. We must keep in the front of our minds that it was NOT the people on welfare who designed the welfare system. These people have merely become virtual captives to the system. Their behavior is controlled; they have little ability to even move to a place with more opportunity. If we take away the Incentive, if we take away the motive, then what is the real purpose of doing something? If we guarantee even a very modest handout (or "welfare" or "entitlement" or "Section 8", or "Aid to Dependent Children", etc.) then we eliminate the incentive, and guess what? Why work when it's handed to you? Why not just have more babies to gain more entitlement? Well, it works, and it works too well, expanding our welfare rolls in a geometric fashion. This seems to have doubled our welfare rolls about every 15 years.

"Here, we'll give you money every month, and the more babies you have, the more money we'll give you". "Just vote for us, and we'll never cut the money off". It was those rich people that robbed you of your heritage and comfortable life after all, wasn't it? You deserve satellite TV; a cell phone, good food to eat and a nice car don't you? Wasn't it your ancestors that made all this wealth possible? Did that same level of wealth exist in the country their ancestors left when they came here? This same statement is true whether people came here from Africa, Europe, South America, Asia, or anywhere else in the world. People built this country by leaving places that were having rough times or were oppressive, came into this country with permission of the US Citizens, and their families grew with our country's growth. If productivity, the actual Production of Goods, did not happen, there would be little for anyone, and no incentive to come here.

It's likely that most people on long-term welfare have been convinced that they deserve free money, food, housing, and more. People on Welfare are made to be virtually invisible to "the public", so it is difficult, other than by direct conversation, to determine which people are on welfare. I'm sure there are numbers of people on Welfare, Public Support, or what ever you want to call it, who are appreciative about what society is doing for them. But, I can't remember the most recent time when someone on Welfare was on national TV relating his or her appreciation to the American People. I'm sure I have seen people who are on Welfare on television, but any I believed to be were not espousing their feelings of appreciation and good will. I think most I saw were complaining about the size of their entitlements or wanted new ones.

So, families who fail so miserably as to not be able to provide support should not be hidden away, should not be isolated from public view, should not be made to believe that they have done anything other than fail, and they should be known to those who support them. They should be known to those individuals who give the charity, or if they receive public charity, known to all taxpayers. In this situation, there is an Incentive to not fail. There is an Incentive to Produce and succeed, and to be a contributing member of society, and to not simply accelerate the burden of "dead weight" on Productive people. We all want to be recognized for being successful and contributing to society, not for being failures and burdens. As a society, we do ourselves the greatest of disservice by hiding and not learning from our failures. The worst disservice is wrongly trying to convince those people and families who are failures, that they are not, and that "society" owes them some debt.

About the 2nd, 3rd, 4th, generation "welfare families" - These are people who are for the most part perfectly capable of taking advantage of a free locally funded education through high school, as a "no cost" opportunity for gaining a "society smarts", not "street smarts" education. This increases the possibility of the psychological "affirmation of self worth" necessary to be a productive member of society. Again, we must mention that the average standard of living of a society is directly related to the amount of things made in that society, and how absolutely important it is for all members of that society to be Productive in making Goods. Free Public Education is a step towards making each member of society Productive. We see this importance in the virtually explosive growth in wealth of people in both China and India. They have recognized that the "Service Industry" does not create any real wealth, but "Manufacture of Goods" does.

We have just set a new record in the US. Forty percent, that's right, 40%, or 4 babies out of 10 born in 2008 were born to single moms. A few percent, mind you, are to mothers who can afford to raise their children without public charity. There is too large a proportion of single moms who are producing 4th, 5th, and 6th generation welfare families. This unequivocally proves the poor planning, poor budgetary estimating, poor scheduling of our Welfare system, which by the very nature of its promotion implies that we will "Fix Poverty".

Even redistributing all of "Rich People's" incomes and net worth would at best create a blip in the comfort of our poor, and take away the incentive for the Rich, who are responsible for the wealth of our country. This drags down Production of Goods, and the standard of living of all, including our Poor. Again this begs a consideration of the philosophy of which is more important: - All income groups of people having a higher standard of living, or those now enjoying a high standard of living being dragged down closer to the poor? Isn't this really giving the Deadly Sin of Envy credence in our Society? Why not allow those highly Productive People a wider berth to help create the opportunities to raise the standard of all, regardless of the relative difference in wealth it causes?

If we look at the differences in relative wealth between our richest and poorest, there is indeed a large spread. However, the "average", which is really the same as the "average standard of living" in our society, has consistently risen over decades and centuries. This has been due to the increases that have been made in Industry and Production. How well did the richest and poorest live in previous centuries, before industrialization?

We have a huge resource of untapped talent. This resource should be utilized. Creating and nourishing a system whereby this talent is forever wasted is shameful. Our real goal should be to elevate the condition of the Poor by giving them the opportunity to have meaningful, Productive work they can enjoy and be proud of. This is what will break the curse of the feelings of worthlessness and shame in the public view. Of course we can train these people to show an external posture of worth, but the innate self-degrading effect cannot be so easily cast away. We can convince people they are victims, but not prevent them from evaluating the sum total of the lack of their own personal achievements and the low esteem this causes.

"When the people find that they can vote themselves money, that will herald the end of the republic." - Benjamin Franklin

SOCIAL SECURITY

Elderly people, the medically unfit, and abandoned children are the real purpose of the "Social Security" system. The purpose and intention of the Social Security system is not to offer a windfall to new arrivals to our shores, or as an incentive to have babies. It is for those who have already put their fair share into the Social Security Trust. Due to the lack of some individuals' long term financial planning, in conjunction with the inability of small businesses to provide a retirement program, the government decided to create an automatic retirement plan. If Social Security went exclusively to retirement payments to those who paid into the system, the Social Security system would continue to have adequate funding into the distant future, as each person and company actually pays into "The Social Security Trust Fund" for their future retirement. We are led to believe that present workers are paying the retirement of previous workers, but in fact those collecting Social Security already paid for their own retirements. If the Social Security System is drained of funds for other reasons, then present workers' contributions must be used for present payments of past retirees, which is not the founding and continuing mission of the Social Security System. People paid into the system for their own retirement. Period. This is a good system, and it used to work well. If the collected funds earned even simple interest, massive funding would be available.

However, a significant amount of the money paid into Social Security has been directed elsewhere. Go look this information up for yourself. Don't let someone else tell you about it - they may not be well informed, or maybe they want to influence you for their benefit, to support their hidden agenda. As of 2005, the percentage of recipients of Social Security benefits to those 65 or older has dropped to an astounding low of 28.8%. Of course there are not less retired people on Social Security, but more. The portion of recipients less than 65 has grown phenomenally faster than growth of the population. Those recipients less than 65 include those for survivor's benefits, medical disability, and a few others. The Social Security System was not designed to be a Welfare system, but it has certainly been warped into one throughout a history of fits and starts.

Based on the same logic, the US Government should supply healthcare to all regardless of the "ability to pay". (And of course many think they will get the same sort of care that Ted Kennedy received, but in an emerging world of the "few special people" and "the peasants", it's not likely),

In order to provide "health care for everyone", those who are already paying for health care will then have to pay their own, PLUS the care for those who are not now paying for it. So, the Productive person's cost for health care jumps up considerably. It has to increase. Who else is going to pay for it? Government money is after all just taxpayer money, isn't it? People who don't pay taxes are now getting a larger and larger portion of the taxpayers' money (and housing, food, heat, transportation, free cell phones, etc., etc.)

WHAT'S GOING ON ELSEWHERE

Some European Countries adopted all-inclusive types of welfare policies many years ago. This did not eliminate crime, did not help the children, and is now bankrupting those countries. Who should bail out those countries? The United Nations? Who does the United Nations bilk for such ultra-massive bailouts? Now we're talking about much bigger numbers than mere trillions. Some believe this is one hidden agenda of $CO_2$ taxation. Fuel combustion is the result of industry that belongs to the productive. Grab the money from those rich bastards. Take rich Johnny's nice red wagon. Johnny can get another one.

As a profound example, Japan was destitute after WWII (the

33

Pacific War), and with few natural resources, a destroyed infrastructure and Industry, a population less than half of the US, and with usable land 1/3 the size of the state of California, Japan built itself up to being one of the foremost leaders in the modern world - within a single generation, all without charity, and all through Productivity. This is the Model of Good Work. The Model of Good Planning. The Model of Success, and an inspiration for the World.

South Korea is following this model. North Korea is not. Singapore is another fine example of careful planning, Incentive and Productivity, as they have also produced themselves up out of poverty over the past 50 years.

Many in our country would have us follow the example of other countries. Unfortunately, the countries that they would have us follow are often the countries who have not been successful. They would have us mimic the Canadian Health system or European economic system or the state of California's social programs. Isn't it foolish to emulate a failing system or policy?

POPULATION CONTROL & WELFARE

Nature puts controls on populations of animals, including humans. The natural biological imperative is to have as many offspring as the food and available space will support. Natural controls on the population of a species are lack of access to food, disease, and social pressures that cause fighting. Human populations are also controlled by these same natural forces. Epidemics, drought, and wars naturally control populations.

A wild antelope does not have babies if there is not adequate food. Some animal mothers actually withhold birth until food becomes available. These natural controls on population are far superior to overpopulation and starvation, ravages of disease and conflict for available resources. How do incentives for having many children cause a problem? Let's make a simple, crass comparison with raising animals. Give them all the food they need, and constantly eliminate the diseases which invariably invade high populations, and guess what? The population explodes. Let's see, that's 2 rabbits, then 20, then 1,000, then 10,000, then 10 million! Puts a bit of a burden on those feeding and providing good health for all those rabbits pretty quickly, doesn't it?

Interestingly, with Humans, another controlling force, that of the tremendous power of the human mind, is Willful Restraint,

34

along with technology. Willful Restraint and knowledge and use of technology are used in most developed countries. Underdeveloped Countries are characterized by having more people than can be supported, as is true in many large cities. The population expands to the point that resources and money must be pumped into many cities from "non-cities", just to keep the people there fed and sheltered, yet their populations continue to explode. This condition is appropriately termed "Unsustainability".

Some societies and religions promote having as many children as possible. In many countries there is no way to save money, so children are an investment. The kids can support the parents in their retirement. No saving of money is possible or is needed. This is the case in underdeveloped places. In the developed world, with equal education for females and males, birth rates are at a sustainable rate, and are actually declining in some highly developed societies.

In a society where only present and future Productivity is valued (where there is no way to save for one's own future, and wealth is not possible to accumulate), a couple with many children is more likely to be supported in their old age by their children, than is a couple with less children. So more is better. 15 children are better than 10, which is better than 5, which is better than 2.1 children per family.

The same is true for farmers. Having lots of children to work the farm (without all of the employer/employee headaches and extra taxes) will guarantee a better retirement for the farming couple, who can then in turn pass the value of the farm on to their children. Having adequate food and health care will not inspire the people caught in a seemingly endless drought to magically cap their explosion in population at a point where the country's resources can support their people. Politicians and decision-makers and charitable groups seem too often to conveniently neglect this hard reality, the predictable outcome of their actions, but perhaps it insures the need for their mission into the distant future.

There has been reported to be a country that is afflicted with a seemingly continuous lack of good crops due to drought - Not just a single event, but continuing over a long time. By receiving gifts of food and medical care, the population has doubled in a matter of roughly a decade. This sounds like a wonderful act of selfless charity and donation. At what point, at what level of population growth do we stop giving? At what point will that society be self-supporting? Implicit in the plea for help is the understanding that once these people realize that a large percentage of their babies will survive, they will somehow reverse the belief that a mother must have as many babies as possible so that some will survive. Does it ever happen? Doesn't the couple with many more children than their neighbors always have a better retirement? More babies surviving does not convince a mother to stop having babies. Education of women in Western Societies does decrease the number of babies conceived. Education of girls, particularly before they start their families, is a critical need for all societies.

If a group saved 99% of the world's children from starvation and disease, without considering other factors, global population in 3rd world countries would uncontrollably explode. A more sensible approach would be to implement an allied industrial development program, birth control, and direction to a 5-year plan for self-dependency. Just like an engineering project with a definitive goal, budget, start and finish times, these social improvement programs must also be planned, scrutinized, held to timetables, and closed according to plan. Or else, it NEVER ends, until "Other People" run out of money.

So what is the important part? Helping a society become self-sufficient, or creating a structure of never-ending and ever-increasing dependency and charity? When does this mission become so important that governments pillage their own people for some other reason ("save the earth", for instance), to dump massive funds in the name of a justified emotion? Missions of governments and charitable institutions should, in addition to expounding their guilt-provoking emotional tirades, explain their path, predicted outcomes, goal, and end-of-mission schedule.

RELIGIOUS AND SOCIAL FACTORS

An interesting phenomenon of many religions is the belief that within a religion "more members is better", for the power of that religion, and belief in the anecdotal evidence that "if everyone in the world believes in the same religion, that the condition of humanity will be 'better'". This is an un-attainable and warped ideal. Its actual goal is the gain of power and control, as these religions often believe and even state that in general, "people are too stupid to think for themselves, and we need to think for them".

At one time in ancient history, Catholics promoted having families as large as possible regardless of the resources available, but they have fortunately realized the ignorance of this action and that strategy for world domination. Some other religions have not yet quite caught on. We can see that in countries in which highly reproductive people are the majority, poverty is generally rampant.

The idea that a society or a government would encourage a high count of births just for the sake of lots of babies no matter who pays, is just absolutely wrongfully absurd. Even programs that "unintentionally" promote this activity are wrong, and MUST be changed. It must be changed so that girls are NOT encouraged to have babies they cannot support by themselves, or can't be supported by their family.

It is a personal decision to have children, and more importantly, to take the responsibility for making the decision to have children and to ensure that those children will also pay the same taxes in their Productive lives, as the parents should now. The Holy Bible says that people must take personal responsibility for their own actions. Deciding to have a baby is an action, and the responsibility taken is to provide for the child until the child becomes productive. This is very simple, and deciding to not have a child is virtually without cost.

Recent studies of the numbers of undernourished people in the United Nations designated "Least Developed Countries" predict a bleak picture of the future. Despite international humanitarian efforts and improvements to food production capabilities, the number of "undernourished" rose from 212 million (1990-1992) to 252 million (2006-2008), a gain of 19%.

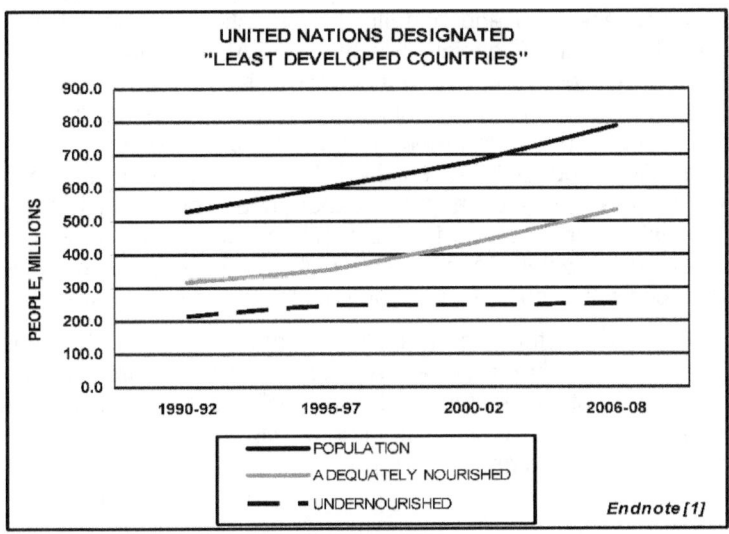

Here, the food supply must have improved greatly, since in the same time span the percent of "nourished" people improved by 68%. Quite an accomplishment, and quite an improvement.

So, if the population had been controlled, or even reduced over this 20-year period, virtually all people could have been fed. In the 2006-2008 period these countries were adequately feeding 535 million people. The population of these countries in the 1990-1992 period was 530 million people. Zero population growth would have solved the problem. Unfortunately, the populations grew during this time span by 49%! (The United States experienced growth of less than 15% of native born citizens during this time.) So the answer for the "least developed" nations is not how many Billions that can be pumped into those countries to further their population explosions, but rather how those countries can do the moral thing and control their populations to be within the number they can support by their own productivity.

When will they suddenly come to the unmistakable conclusion that since as many people are starving to death as are being born, that it will be time to do something? Or will the breadbaskets of the world possibly be able to ship in enough food and medicine? Are we talking food wars and water wars? Massive pestilence and disease, plagues on a scale the world has never seen? All the developed countries of the world, giving the sum total of their remaining combined wealth could then hardly make a dent.

China has made the strategic decision that population control is essential and beneficial. Population in China is now controlled, with financial incentive AGAINST having many children, as contrasted to financial incentives for having MORE children in Welfare, as in this country. Of course children are expensive. The Government is not morally chartered to be financially responsible for children. Parents are, and have over the course of the human experience, been individually responsible for providing for their children. Imploring, forcing, or exploiting a society to be responsible for children is NOT an advancement of society over early hunter/gatherer humans.

China's insistence that families be able to support their own children is exactly the correct and difficult decision that modern societies must make. Otherwise, a Society's failure is virtually guaranteed. China as a country should win a special Nobel Prize for this accomplishment.

EDUCATION

We give all Legal Americans the right to a locally funded public education, which is paid for by local landowners through property taxes. It is up to each person to make use of that right, the value of which must be taught by the young person's family. In the same vein, if a person gave you some seeds to plant, you could plant them and literally enjoy the fruits of your efforts, or you could eat the seeds, or just throw them in a dump.

If an individual or his parents do not take an educational opportunity, then shame on them. It is up to the Local City or County government, and to the people living in that city or county, to correctly insure their program for public education is good for their people. It is NOT up to the people of Dubuque, Iowa to provide resources to the people of Billings, Montana. It's up to Billings. It's not up to the State of Montana; it's up to Billings. This is not a responsibility of the American Citizens; it's that of the people of Billings. It's up to the people of Billings to make sure that proper emphasis is placed on their educational system and supply the funding for that system.

**A free public education is a right to the opportunity to learn and to become a Productive American. This is the limit of, and the totality of "Economic Equality".**

Public education, in concert with family teachings of morality and fair play, in personal responsibility and in good work ethic, are, in all but the rarest of instances, adequate to prepare a youth for a lifetime of Productive Work, providing for the wealth not of a government, but of the individuals of a nation.

WORD OF WARNING

There are those who have their own agenda, who benefit from the control of financial resources and their self-serving dictates to Educational Systems, to the point of directing the composition and morals taught there.

THE RACE CARD

During our Country's early development, ending now almost 150 years ago, people from Africa were forced to come here against their will, but since, have added immeasurably to the strength and prosperity of our country. Numerous Americans of African descent have achieved commendable contributions to society and to therefore enjoy commensurate riches.

The growth and riches of "The Old South", related to us through our history books and in stories such as "<u>Gone With the</u>

Wind", were facilitated to a great degree through Greed and Exploitation of Slave Labor. The Civil War, a dramatic turning point in American History, cost virtually all of the wealth of The Old South. Riches in the North, on the other hand, were morally earned by the Productivity of invention, machines, and the Industrial Revolution. It was the expenditure of a huge amount of the riches of the North, which ultimately lead to the defeat of the Confederacy, and the end of Slavery in America. Many believe the psychological wounds and social stigma over Slavery continue until today, and Americans of African descent must continue to work harder than those of say European descent, to earn the same standard of living, and to earn the same position of admiration in our society. This "Black Tax" comes from a sense of having to overturn what some feel is a history of lower societal worth, which has of course been proven to be in error. This is a mental wound now proven to be easily and totally curable through Opportunity, Incentive and Reward.

The present Wealth of our Country is totally independent of previous work. None of the people who suffered from or benefited from Slavery are alive today. Many examples of people who came from Poverty and who have ethically created personal wealth totally dispel any theories that what happened in our distant history has any bearing on the situation today, except as some sort of mental wound, which Opportunity can cure. This is much easier when we ignore the illogical demands of a few, who would keep those wounds open and festering for their own ulterior motives and benefit.

Some came here not of their own decision. That was a crime, and those victims and the criminals responsible for it are all gone. No one alive today in America is guilty of any step of the Slave process. No one alive today in America has any of the "spoils" of those deeds of immoral greed. "The South" blew all their money on trying to maintain Slavery. "The North" spent tremendous resources and lives to end it.

IMMIGRATION

The People of the United States, Americans, are a diverse mix from all over the world. The original Americans from Mongolia and Scandinavia had the same incentives and motivations to better themselves, to move to a better place, as have most of our volunteer Immigrants. Since the US was established in 1776, the American People, through our government, have selectively allowed a carefully controlled mix of people to enter and eventually become citizens.

Our forefathers and foremothers engendered the characteristics that have allowed us to grow in strength and prosperity over the years. A keystone of that spirit of development is allowing entry of "disease free", legally clear, non-criminal individuals. In fact, during the oversight by "Ellis Island", 98% of all applicants were allowed to enter this country. Many of these people decided to stay in America.

The descendants of virtually all who have come here with our permission now have equal access to the cornerstone of Individual Productivity and therefore the future success of our country. That is, a good, basic, locally funded public education. A free education through high school is offered to all Americans regardless of their "position of birth". Individuals with exceptional capability to contribute may continue their education, and make tremendous Productive contributions by making more things available to more people, and therefore deserve to enjoy proportionally more of "the good things" in life. These extraordinarily contributing people should not be discriminated against, and all must be treated equally under the law.

People must not be discriminated against based on financial standing, nor their DNA, creed, or other basis. Our laws must be complied with.

Americans are people born here of parents legally residing in the US at the time of birth, and people who have become naturalized through the traditional process. Putting in bluntly, a child born to "Illegal Aliens" is not an American citizen, and neither the child nor the parents are entitled to any handouts or free education, only a free ride back to the border from whence they came. The Fourteenth Amendment to our Constitution applies only to American Citizens, as does the entirety of our Constitution. [2]

At the time of its creation, our Law to mandate that persons born on these soils had the opportunity to choose to be American when they became of legal age and illegals were not permitted to enter, much less remain here. Therefore, the automatic citizenship law does not apply to children of illegal aliens.

Virtually all countries have similar or even more stringent rules regarding "uninvited" immigration, and they continue to aggressively enforce those rules. We've somehow placed less emphasis on these laws, not due to the Will of the American People, but rather the will of a few with their own greedy rationale. These rationale undoubtedly include cheap labor by people easily dominated and the Hope for Change in Voter composition. The number on non-native residents in our population has exploded from approximately 21 million in 1991 to 38 million in 2007. This represents a growth of 81%, AND does not include the offspring of these individuals who have been born in the U.S.

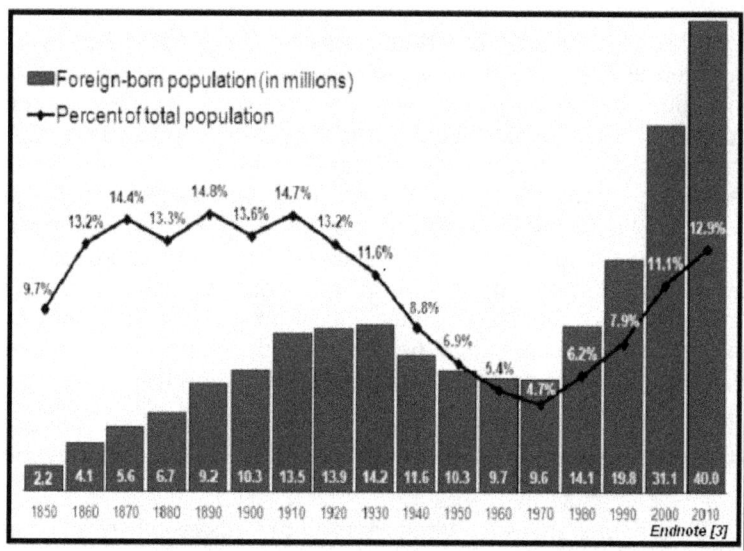

We in the US reserve the right, as a Judeo-Christian Nation, to maintain the standards of morality of the "Ten Commandments". We are not to kill, steal, etc. We maintain the protection of individuals against immorality, of sexual abuse, of forced advantage over others, and believe in the concept of fair play. Other countries do have morals significantly in conflict with ours, but if people wish to live here, they must comply with our laws and morals. We cannot, of course, force our morals upon those in their countries, and in turn, those countries cannot vilify us for our own morals and preferences. If a society believes that subjugation of women is their heritage and their right, it is their right to practice this subjugation, whether we like it or not. Likewise, It is our right to practice equality of the sexes.

When our uninvited visitors ("illegal aliens" as written in our Governmental Records), return to their own countries, many opportunities will be opened for Americans now on Welfare. The life of a Migrant Worker is after all honorable, full of healthy exercise and outdoor work, but of course doesn't pay $100 per hour. The physical condition of Migrant Workers actually approaches the ideal. Few are fat, most meals are healthful, and all that is needed to make it an ideal occupation is having an accessible infirmary and a directed savings plan. This would break the bonds of dependency, promote Freedom and the opportunity for a self-directed and Independent path of life. It would allow the joy of being a contributing member of society and the claim to be a part of Our Country's successes.

CONCLUSION

Is a farmer who is struggling or even succeeding in providing for his/her family also responsible for others who have made wrong decisions in their lives? Is he/she responsible for those who have shrugged off their opportunities for education, for work, for bettering themselves?

Incentive is the thing that drives everything in the "kingdom of life", including humans. That incentive being the needs of "The Basics of Life" (food and shelter), position in life (status), and the wherewithal required to support one's offspring.

One philosophy about social structure is that all people except the ruling class should be equally poor...this is of course perfect equality, (except for the ruling class). Human history abounds with cases in which this strategy has failed, but for some reason it persists in backward countries. There is the Leader and their entourage, and everyone else is a peasant. America seems to be moving in this direction at the present. We're preparing for massive increases in costs of Energy, changes against Industry, and there will necessarily be far fewer truly Productive Jobs, less Production, and therefore less goods for everyone as a whole. We cannot make up for loss of Productivity by engorging the size of Government, thereby increasing the total number middlemen between people and services, no matter what those services are.

So our Social Responsibility is to treat all others as equals, and equally respect the property of others. It is to insure the future improvement of our society for all with our local public K-12 school systems. It is our responsibility to ensure that this education includes our society's history, reading and writing, mathematics, and also the lessons of morality taught by Family, Church, and the Constitution.

Far too common, although still devious, public charity is used to sway votes for candidates associated with a particular political Party. In this form, money is taken from those who have earned it and deserve it in order to buy votes from others. Those same others, who are likely programmed by that same party and their operatives are often willing recipients of that Charity, and feel they owe their support to that political party.

Individual citizens may give of their own personal charity, from their own hearts, by their own free will, to any cause or condition they desire (Not charity for political upheaval through funding terrorism, or other antisocial purposes, but by the heartfelt concern for elimination of poverty and disease, or improvement in the arts or other cultural areas). This charity may be to voluntarily supply food and medical care for short-term relief, and birth control for long term relief. The point of birth control must not be underestimated. If aid organizations and charities had encouraged population controls 40 years ago, there would be little poverty in the world today. Wouldn't that be a wonderful condition of humankind! Isn't this the actual, real mission we should be promoting for the Global Society today? The huge detractors are the governments and institutions who want to maintain poverty to continue their control and missions of today.

Many touching stories of individual charity to the cause of children abound, stories where the parents openly show their gratitude, and do not demand that "Society" owes them yet more.

Even Mother Teresa knew the limits of the progress that could be made in a segment of a society. She did not believe "the rich" could be forced to save "the poor", but that the poor could save themselves and develop their own sense of pride through development of their own community and their own industry. Morally, wealth is not taken and redistributed, it is earned.

# MONEY

## REAL MONEY

Very simply, value is created only through making goods. Money only represents goods made. Today, in America, we do not make as many goods as we used to, and that is why there is less to go around. We have inflation as more money is printed than represents the additional goods made in each subsequent year. This has the same effect as counterfeiting. The theft or destruction of something of another person is in fact, and in deed, taking away a part of that person's life, as is the case with many of our Taxes. The same is true for inflation of money. Loss of the "buying power" of money a person has previously earned and saved is a loss of a part of their life to inflation.

Money is an exchange medium, and that's all it is. Money represents tangible goods, for example: a fish, a pound of rice, a ton of iron ore, a cabinet, or a bottle of aspirin. Early money, and some money today, has an intrinsic base value due to its significant composition of gold, silver or platinum, or even salt. The denomination of the coin is proportional to the weight of the valuable content. As a convenience to both governments and people who carry money, representative notes are printed, which state an exact weight and exchange rate of precious content held in a secure place to equal the printed notes, (for instance US "Gold Certificates" and "Silver Certificates"). Somewhat less confidence exists for certificates than actual metals, but trust in proven integrity gradually creates a similar sense of security in precious metal certificates and the durability of the value over time.

Holding money that has the intrinsic value of precious metals creates a sense of continuing un-changing value.

For instance, if a baker has baked and sold a loaf of bread, the money the baker received is spent to buy flour and shortening, as well as other needs, and some money is saved. The saved money has the same value, and can buy the same amount of other goods either immediately, or for many years into the future. This is the real meaning of money. It is a way to store the value of goods and labors for future use, and to avoid the inconvenience of only trading goods for other goods. In other words, money is a placeholder and is representative of goods, both "durable" and "non-durable". Money only represents goods, as it always has, and can be exchanged for goods or services, at market value.

Money is issued by "authorized" authorities. Costs of providing both precious metal coins and their corresponding certificates are very nearly equal to the value of the denomination of the money. In these cases, the government authority is acting as a non-profit agent to facilitate commerce and growth of specialization in society, thereby, and in the words of the Preamble to the US Constitution, to "promote the general Welfare". With money as a means of exchanging value, commerce with distant places is enabled, and each person does not have to catch their own fish, raise their own animals, or make their own clothes and shelter, etc.

So the purpose of money is merely to facilitate an exchange of goods or services for goods or services. It is exceedingly difficult to abuse a system of money based on enduring value. The awkward times for real value money as herein described occur when there is a change in availability of the precious metals, particularly if the total amount of available metal becomes excessive, or rare. For instance, if everyone had a few tons of gold, its relative value would diminish.

FUNNY MONEY AND THE HIDDEN TAX

We get into trouble when money moves to a non-intrinsic value (not exchangeable for a certain amount of precious metal, for instance). This allows the government to manipulate the value, and is usually at a loss to the people and for benefit of the government. Many have heard frustrating stories of taking a wheelbarrow of money to buy a loaf of bread in Germany towards the end of the Reich. Their money was not enduring. "Inflation" or loss of value of that money was extreme. Another example is money of the Confederacy during the US Civil war. In those times/places, someone earned a certain amount of money, and had to immediately buy something, anything, as their money had the terrible affliction of evaporation, like a chunk of ice in the desert. Not even water remains after a very short time.

Many nations now have money that is based on a promise, such as our "Federal Reserve Note". The un-spoken promise is that the value will remain constant. In some societies, that promise has been broken too many times by demanding and overpowering governments, and in those places, only money based on precious metal is now accepted in trade. Typically, these same societies have a dualistic money system, accepting the promise notes and immediately exchanging them into silver, gold or platinum coinage for savings. Even in the U.S., although inflation is not yet universally realized to be a tax by the government, money earned in 1913 has already been taxed 96.9%, due to inflation alone. Inflation is a tax, every year, of ALL money that has ever been saved. Money saved from earnings in the year 2000 has already been taxed over 25% through inflation. That loss of value has NOT been transferred to Capitalists. It has not benefited "The Free Market". It only benefited the Government, and the creation of too much money caused it.

Stopping inflation, stopping the loss of the value of money is the only way to preserve the value of our work. Some believe that "Deflation", that is an increase in the buying power of each dollar, is not good for an Economy. Interestingly, however, Deflation benefits every person in a society. Everyone can buy more goods with the money they have, and the money they earn.

Natural deflation, the reduction in the cost of goods enabled by lower raw materials costs and improvements in productivity is good and is healthy. Price and actual cost of electronics goods is a perfect example of healthy Deflation. As most governments maintain a policy of planned Inflation, many occurrences of Deflation are during times of economic depression, so our image of Deflation is associated with Depression. A Great Depression is bad. Deflation is good.

Let's look at two scenarios from a governmental perspective. In the first, precious metal money, or certificates for an equal quantity of precious metal are examined. At one time the US Government kept a troy ounce of gold in Fort Knox, Kentucky for each $20 (later $32) it issued in gold certificates. The cost of producing the certificates and safeguarding the gold was (just a guess, research it if you want), roughly 99 cents for the precious metal, and one penny for the printing to make a dollar bill. We then had to be taxed one cent for the paper dollar bill, which might last for 5 years or more. After the 5 years, one more cent was needed to print a replacement bill for the same dollar. Of course larger bills, such as a one hundred-dollar bill also cost only one cent to print. Large bills, as they are not physically traded as often as one dollar notes, can then last 10, 20, or more years. Still, the true value was "backed" by the correct amount of precious metal, remaining in the federal repository.

The task of the government was therefore non-profit, and truly met the mission of providing a currency used in trading goods/services for goods/services. This is one of the few actual missions of a benign government. - To standardize the money without making a profit on it.

In the second scenario, with money, for instance a Federal Reserve Note, promising enduring value, it no longer costs one dollar to produce a dollar bill. It only costs one cent. In this case, the government can print 10 bills of $100 denomination, and the government can make a "profit" of $999.90, without needing to purchase precious metals, (or provide a storehouse and staff to hold them.)

Inflation is the loss of value of money. In a "pure" system, if a certain amount of goods is produced in a year, let's say 100 bushels of corn, the amount of value of this production is represented by $200 at 2 dollars per bushel. So, the government must insure there is a supply of $200 of money in the system to allow the trade of the 100 bushels of corn for other real goods. This is not an addition of $200 more dollars, but just making sure there is $200 available already in circulation. Some of the money may be spent for a hair cut or lawn service, or tractor parts and fuel, but the basic value of the corn and money is unchanged over time, and there is no inflation.

In some years it may be an excellent growing season, and a farmer may make 150 bushels of corn per acre, as may all other farmers, and the cost of producing the greater amount of corn is roughly the same as the 100 bushels the year before, then the market price of the corn, due to the supply and demand forces in a free market, may be lower than $2 per bushel. In other years normal and natural variation in weather may curse the farmers and allow only 50 bushels of production per acre. Then, the price will be higher for each bushel. But basically, and for the purposes of this discussion, the intrinsic value of the corn is constant, as is the value of the money printed to allow the exchange of the corn for other goods. And this is fair, as the value of the use of the corn and the true value of the money doesn't change. That same $200 will circulate through society, and in fact the same bills may be paid back to the farmer the next year for another 100 bushels of corn. So, the government does not need to print any more money as long as the total of all goods produced in a year is constant.

If the government prints more money than required for circulation for all of the corn grown, cars built, and other production, then automatically the price in dollars goes up for each of those things. We call this inflation - basically more money is chasing the same goods.

The "Monetary policy" of a government controls the amount of money available, supposedly promising to keep long-term value and prices un-changing. In fact, without inflation, a person can save and store money in their own safe place, under their pillow or wherever, and it will buy the same amount of goods if spent immediately, or if spent 50 years hence. In this case, people have an option to "get by on less' (a.k.a. "live within their means"), if they so choose, and save the remainder for their future security, knowing that they can take care of themselves in the long term. They would be accepting responsibility for themselves, and if they want, may at some time retire.

Inflation comes from putting too much money in circulation relative to the amount of goods produced - The larger the amount of money compared to goods produced the higher the amount of inflation. The benefit to the government for producing more money than necessary is that the government can then get out of the non-profit mode. Every single extra dollar printed allows the government to buy an extra dollar's worth of something, and not depend solely on regular taxes for its income. The sinister side of printing the extra dollars is inflation, which compounds over time. If, for instance the government injects an extra 5% of dollars each year into circulation, then the value of the dollar decreases 5% each year. The dollar cost of each bushel of corn goes up 5% each year. Then, the government has in fact levied an extra 5% in taxes, without raising the apparent tax rate.

This inflation is not only a tax of 5% on the farmer's income for that year, but also a tax on every single dollar the farmer has saved in previous years. At a 5% inflation rate, half of the farmer's savings are gone in less than 15 years. If the farmer has scrimped and saved, and worked as smart and hard as possible, and was able to save $100,000, the tax on the savings alone is $5,000 per year. This monetary policy of 5% inflation is a financial attack on the farmer, depriving the farmer of the toil and sincere discipline of her/his work.

With the farmer's money now in a bank earning some interest, probably not as much interest as the inflation, there is a "hope and a promise" that the original value can be maintained. The farmer is not trying to get rich on the savings; he's just trying to maintain the value into the future. Of course with modern information technologies and rules and regulations for the banks, the government is free to continuously monitor the amount of money the farmer has, and what the farmer does with his money. Perhaps the motivation of maintaining observation of the farmer's money is to protect others from possible criminal activity of the farmer. "Anti-money laundering" is probably as good a rationale as might be claimed for this monitoring. People have in the past, in order to protect their money's value as well as maintain the privacy of their business, put money in foreign banks, or bought gold or other durable goods.

As good a rationale as we could apply to forcing the farmer to put money in a bank, at no actual gain for the farmer, is to make money available for other people to borrow, to build their business, or buy their home. If those other people had however, been good hard workers and saved a portion of their earnings, they would already have their own money, and would not need to borrow anything. If the other person wanted to get money to start a new business, they could bring some of their friends or some other investors into their business, and start the new business. Interest payments would not then be an overwhelming business expense, allowing the business owners to focus their attentions and efforts more to the actual business of providing more goods (and services) of better quality at a lower price, while having a higher percentage of employees focusing their talents on Productive work, rather than non-Productive financial matters.

So, in the instance of the farmer putting the same $100,000 he saved over countless years of back-breaking work into a savings account with a 5% interest rate, during times with 5% inflation, it would seem the farmer has achieved the objective of protecting his money. But no, he hasn't. The farmer must pay taxes on the interest earned, (if the farmer is one of the 50% of American "workers" who pay Federal Income Taxes, that is). Morally speaking, the farmer should be allowed to deduct the inflation of the amount of their savings from the interest earned, as the farmer did not, and could not actually earn any value from the investment, due to inflation.

We need look a bit deeper here into the concept of "inflation", which we are now programmed to tolerate, to accept as being a normal part of a Free Market, Capitalistic Society. We are told by "the left" and even some economists, that Inflation is the result of greedy Capitalists. When we go to a store to buy something, we see a higher price, and the storeowner can only shrug his shoulders, and explain that the wholesale cost of goods is increasing. The wholesaler claims the same of the distributors, who claim the same of the manufacturer. We speak with the factory management, and they speak about the increasing personnel expenses, the cost of energy, and the cost of raw materials. So, who is to blame? No one in this sacred supply chain is guilty of greed. Each has an incentive to supply more higher quality goods at lower prices than his competitors, which is a core facet of our successful system. Free market forces of "Supply and Demand" prevent any Supplier from charging too high a price, as someone else will sell at a lower price, taking away part of their business and perhaps expanding the market.

Here's an example that should bring most people to a correct, logical understanding. It's been said that Americans have a total "net worth" of about 50 Trillion dollars. (Of course some officials are trying to figure how to take a lot of this net worth, and give it to people who have made poor "life choices", but Personal Responsibility is beside the point in this discussion.)

In determining the total net worth of Americans, totally ignoring the immorality of taking the wealth of Americans and spreading it around, we must realize that the computation of total net worth is simply book value. Simply put, if 80 percent of total net worth is in the form of homes, stocks, and durable goods, and 20 percent is cash and certificates of deposit, the actual total net worth is much lower. Why? Because to determine the value, the 80% of things must be sold. The first house sells at book value, but the very last one is worth nothing, and can only be given away. The first shares of stock in a company are sold at book value, but as that stock continues to be sold, the price drops, and the last shares are worthless. The money continues to have its original value, less inflation. So "Total Net Worth of Americans" is not 50 trillion, but is in fact closer to 30 trillion.

As the Government well knows, a 5% inflation in the value of money, (a dollar being worth a dollar, to a dollar being worth 95 cents), brings in a tremendous income to the Government. Let's only consider 5 trillion in "cash" savings (15% of the Total Net Worth), 5% inflation of 5 trillion is 250 billion dollars extra Government "income"! If we consider also the "investments", which of course lose value with inflation, then it's $1.25 Trillion extra income. This is totally obscene, immoral and unfair.

The Government 'prints' an extra $1.25 Trillion, and people therefore lose $1.25 Trillion.

There is a strong need for some Governments to make people believe inflation is solely the result and fault of "The Free Market System". By making more dollars than needed they are in effect acting the same as those who counterfeit money. The Government is thusly diluting and collecting from the savings of all. The Government is collecting from those who plan for their futures, and who save their money. Inflation IS a tax on personal savings and investments.

When we look at motivation for the immoral monetary policy of continuous inflation, there is firstly the extra tax on income and net worth, previously mentioned.

Secondly, there seems to be a voyeuristic desire by governments to know how much money a person has, when it is used for something, and what it is used for. If people could save money themselves with confidence of its future value, they could hide it in their mattress or a wall safe, or in a jar buried somewhere. Then, the farmer or a descendant could one day use it for something. There is certainly no immorality in this, the farmer later using his/her own money, or the family using this money for some purpose, such as buying a new tractor, installing a silo, or maybe a nice new saddle for a grandchild's pony. With inflation, the farmer is forced to "invest" it somewhere, to try, really, to offset inflation, to maintain the value of their historic efforts.

Inflation is not secret, nor is it intended to be so.

In defense of the new round of Inflation which will be caused by today's new round of printing more money, the "Experts" are saying that Deflation is bad. As previously stated, Deflation is associated with "The Great Depression". This is not hard to understand. The Great Depression caused Deflation. No one had money to buy stuff, so the people selling stuff had to reduce the price. No surprises here. The Free Market in action. Deflation DID NOT cause The Great Depression, but was rather just an effect of the natural market forces that are always at work.

When we can reduce the costs of Production, reduce the costs of raw materials, increase the efficient use of raw materials, reduce the costs of labor, the costs of taxes, the legal costs, the cost of energy, the costs of warehousing and distribution and the costs of selling, then the prices of these goods decreases. When this occurs over an entire society, it is Deflation, as the same number of dollars buys more stuff. Competing Suppliers try to sell their goods at lower cost to increase/maintain the amount they sell, and need to reduce their costs to do so. This is not evil; it is not bad for an Economy. This is great for everyone. It's good for the Producers, as they can create more return for their owners. It's good for people with very little money, as they can buy more. It's good for people with all levels of incomes and net worth. So, Deflation is actually good. Only the Government loses the extra income from inflation.

People only want to enjoy that which is earned by their work, not to have it taken away by a bully or a Government.

Here is a graph of the Percent of total tax through Inflation; based on the year that money was saved.

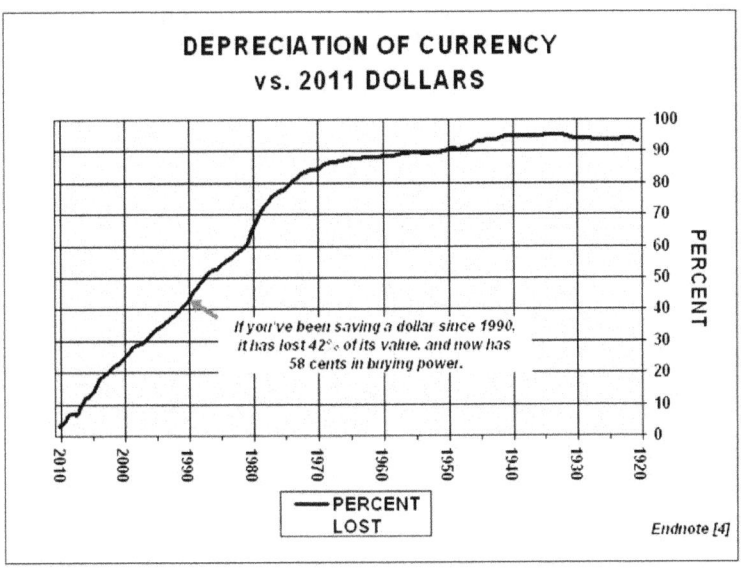

**DEPRECIATION OF CURRENCY vs. 2011 DOLLARS**

If you've been saving a dollar since 1990, it has lost 42% of its value, and now has 58 cents in buying power.

PERCENT LOST

*Endnote [4]*

"If the American people ever allow private banks to control the issue of currency, first by inflation, then by deflation, the banks and corporations that will grow up around them will deprive the people of all property until their children wake up homeless on the continent their fathers conquered." - Thomas Jefferson

(In Jefferson's day there was no Federal Reserve Board, which today has the power he warned against.)

A dollar saved in 1920 lost (was "taxed" at) 90% by the year 2010.The level of Inflation was much lower up until somewhere around the 1960's~1970's, then it increased a lot. This coincides with the time that precious metal backing of the US Money System was halted.

Money saved in 1920 lost about 50% of its value by 1971, over a period of 51 years. Money saved in 1986, by comparison, lost over 50% of its value by 2011, over a period of only 25 years.

Inflation taxation of the American People has increased by 100% from the time before 1960~1970 until the time since. We are the generation that spends more, yet gets less.

So this is money. It is really very simple. Of course "Experts" would want us to think it is very complicated, that we cannot possibly understand it, so that we will accept any proclamation those Experts make without question. Then, the Experts are in control.

# TAXES

Many countries and their citizens are familiar with the phrase "as sure as are death and taxes." Let this not force us to think that we are powerless over taxes as we are over eventual death. We do have a say in the expenditures our Government commits to, those expenditures that undeniably lead to greater taxes, including the inflation tax. The inflation tax is falsely attributed to "those greedy big companies", but just like you and me, those companies struggle to keep their finances "in the good", to stay out of bankruptcy, to improve and grow, and to hire more employees for more Productivity.

Collectively, we support our necessary public servants and our defense. As each of us receives these benefits, each of us has the obligation to contribute equally for this benefit.

Local taxes provide an equal opportunity for education through high school, regardless of the amount of each family's Productivity or income. This is equality, and most believe the equal opportunity for an adequate education to be the core and essence of one of the intents of our country's founding folks - "The freedom to pursue happiness". Previously, only "the rich" could afford to have a basic education provided for their children. This one social program, a basic education, has allowed ample opportunity for everyone willing to apply themselves, regardless of their family's level of income, the freedom to pursue success and happiness.

ROBIN HOOD, THE PATRIOT

Let's consider the following oxymoron, even though this term doesn't adequately describe the degree of confusion, the conundrum of the phrase, a "Robin Hood Government." At first glance (and what socialists would want us to believe), it sounds as if Robin Hood was trying to spread "Economic Equality" - Wherein everyone gets the same of everything, regardless of the degree of their personal contribution to the society. This philosophy is embodied in the phrase: "Taking from the rich, and giving to the poor".

In actuality, Robin Hood was one of the original "Tea Party" advocates. Robin Hood was against high taxes and against overwhelming, oppressive government. Robin Hood didn't take money from a rich citizen. He didn't take from the honest rich; he took from the oppressive government, the Sheriff of Nottingham, who had forcibly taken excessive taxes, "for the people". Robin Hood took the money back, and gave it to the people who earned it, the villagers. Robin Hood did not take the money from a rich merchant or farmer or lawyer, and give it to people who didn't work or contribute to society. Robin Hood was very much like our original American Patriots who rebelled against high taxation.

A notable case of patriotic activity was "The Boston Tea Party" of 1773, being revisited in spirit at this very time. There is a certain degree of vilification of "Tea Party" patriots today among Liberals and Socialists, just as there was against the American Patriots whose actions lead ultimately to our Revolutionary War against our previous Master, Great Britain. The British had a lot to lose and were desperate to maintain their oppressive grasp on the American Colonies, to continue to milk tributes from them - As America's early Spirit proved highly Productive, and was promising to be quite lucrative for the British.

This is a nearly perfect analogy to U.S. taxes of today. The Liberals are getting desperate to keep and claw their way ever so much more so into "America's Pocketbook". Many actions by our present Congress, by our elected officials, is perfectly aimed to dig in deeper, to "squeeze blood out of a turnip", augmenting the process of accusation and condemnation through guilt and shame, using the full spectrum of human emotion, for any and all 'social' causes... Causes for "The Victim".

What does it take for an arm of the socialists ("ACORN®" et al.), to convince people that they are victims? Victims of the very machinery of our society which have made the opportunities and the wealth available to support so many "Victims" of today? Victims not of destiny, but rather Victims of their own choosing, Victims of those who want them to become and to remain "Slaves of Kindness".

These "victims" have been taught to be hateful of their benefactors and to be inappreciative of the "free ride" they have been granted. They have been taught to be angered by the "greed" and "luck" of those rich conservatives who have footed the bill, and to be envious of the possessions and positions of others.

In the place and time of Robin Hood, the non-working people must have consisted of only the old and medically incapable persons, who were cared for by their Families and their Church. There was no tolerance for people who could work, but did not. Everyone found something they could do to contribute. To make baskets, to offer housekeeping services, and to provide manual labor were a few of those jobs. Fortunately, as now, family and charity care were frequently the case - families taking care of the retired members in their family, and taking care of the sick in their family and community. This same spoken and sometimes unspoken "social good" is referred to frequently in religious and other works and in enduring social commentary, through the moral prioritizations of: 1.) family/religion, 2.) community, and 3.) country.

### TAX AND MORE TAX

Almost all money going to a government is a tax. In addition to the items normally called "taxes" are tickets and fines, permits, license fees, federal access charges, and other items too numerous to list. Paying a fee for a government provided service such as water, gas, or electricity by volume used, etc. is simply paying for a commodity, as long as the cost is competitive with similar services in other locales from private suppliers without "a monopoly". These are actually purchases of goods, and not taxes. If water supply from a Governmental agency is more expensive than if supplied by free enterprise, then the extra cost is a Tax, no matter what the rationale.

Another theft of major consideration is the "estate tax". A Government double taxes someone's earnings! Once when the money is earned, and once when the person dies. Of course "Estate Planners" and Lawyers love this law, as it makes a lot of money for them, as they help "preserve assets".

An example of a relatively new tax In Virginia, (as well as in other states), is a "Land, Air, and Speed" program promoted to reduce highway deaths through catching "D.U.I." drivers (who may or may not actually be driving safely), with costs to that driver of many, many thousands of dollars. The first year this program was instituted in Virginia, highway deaths stayed the same as the previous year, or increased a bit and we didn't hear any triumphant proclamations. The second year, halleluya! Highway deaths reduced! We later find out that people are driving fewer miles, and the reduction in deaths is exactly proportional to the reduction In mlles drlven. BUT, mlllions more has been collected by Law Enforcement and associated suppliers. This money doesn't all go to the government, but a lot goes to lawyers and training schools and classes that are in the government's graces. It's a financial punishment to those drivers. But, since there doesn't seem to be any real benefits in the statistics, it's just another form of taxation, in the name of some glorious cause. Maybe this program is forced by Federal mandate to qualify for some sort of funds for highways or another program. Of course if someone is driving erratically and dangerously or aggressively, they should be convicted on that basis, whether they have an actual DUI issue or road rage, if they are incapable of safely driving, etc. After all, it's really about driving skill and safety, and not about personal chemistry. None of us wants injuries or deaths, but we can "look at the numbers", and see that this particular philosophy of enforcement is just expensive to "the public". The "teeth" of the program, the threat that "We're out there", has only seemed to make violators nervous, not change their behavior. Public Education programs are more effective and the cost to our Society is far less than the air and speed types of programs. Common sense thinkers ride with a designated driver, they are not intimidated by "the law" - it's just good common sense. So, this is an example of a tax, a tax "if you're caught", which pays for more people out there "to catch 'em". If we look at the problem which causes MOST highway deaths, it's "Aggressive Driving", whether DUI is involved or not. It's something we see every day, but rarely do we see aggressive drivers caught and get a ticket. It's very likely an "Aggressive Driving" ticket is much cheaper than a "DUI".

PHILOSOPHIES OF TAXATION

Taxes have been collected based on various rationales, and for various purposes. Modern governments have what they consider a "fair" calculation of the amount a person is compelled to pay. These range from percentage taxes upon ownership and leasing of things such as property and vehicles, access to utilities and public transportation, purchases of things, and special additional taxes for certain goods such as alcohol and tobacco. Of course the percentage of these taxes varies widely across the globe, but equalizing these high taxes across all nations would be an absurd act supporting a "global equality" of economies. This is crazy, as public costs are not equal for all nations.

Another basis for taxes is amount of income earned. In a society with "progressive" taxes, earning more money confers to an individual the dubious honor of paying a higher percentage of their income to taxes. Inflation presses more and more people into a higher "bracket". Even when their income just keeps pace with inflation, peoples' actual "buying power" decreases, due to paying a higher tax percentage. This is in addition to the loss in value of their savings and is an additional tax.

Regarding fairness in taxes, we must look first at the cost for each individual, and tax each individual accordingly. Otherwise, we have discrimination of individuals and discrimination of groups of people. Avoiding discrimination is a hallmark of US society's progress. If, for example, the cost per individual of all the real, justifiable national costs, (approved directly by the people, not implied through election of a party or candidate) is $3,000, then each individual owes $3,000 in taxes. This is a "fixed" tax plan. Other usury taxes apply for extra things each individual gets, such as tickets for visiting a National Park. This is the only truly non-discriminatory way to have taxes. It is also so incredibly simple that each person can take care of their own paperwork, freeing up a tremendous number of people (lawyers and accountants) who can then become involved with Productive pursuits.

Direct equality of taxes, everyone paying the same amount of tax per person, whether they work or they are young or old and do not work, can be demonstrated with this simple example: If an economically disadvantaged person goes into a store to buy something, they must pay the same price as anyone else who wants to buy that same thing. There's no "drinking champagne on a beer budget". You pay for what you get, and you get what you pay for. If I purchase a radio, I expect everyone else, whether rich or poor, must pay that same price for the radio. Much in the same vein, Taxes buy things for us through our Government. Since each person gets the same things from the Government, everyone must pay the same amount of money for those goods.

So, the only truly non-discriminatory tax system is this "fixed tax" system.

Another design of a tax system, although discriminatory in theory and in fact, is to take taxes equal to the amount an individual earns in a certain period of time, for instance one month, as their annual tax payment. The discrimination here is that some individuals are more productive, and therefore must pay more money in taxes, and people who are less Productive, or un-Productive pay less in taxes. This is known as a "Flat Tax". One rationale for a Flat Tax system is that every individual then donates an equal amount of time to society, BUT, it discriminates against an entire group of people who are more Productive, and have higher earnings. And under this system there are no "negative taxes", no incentive payments for people to not work and to those who earn little or nothing in a year.

Even in a society with a "flat tax", where everyone, even

those on public charity are compelled to pay 10% of the money they get in total taxes, the highly Productive person still gives more to society. With a Flat Tax, each person is asked to give equally of their time, and this taxing system still honors the "opportunity to excel" which is provided by that society. Progressive taxation, with no taxes for 48% of the people, and the majority of taxes paid by the upper 10% is highly discriminatory. Our Constitution says that all Citizens must be treated equally under the law. One person paying 0% of their income and another paying 50% is not equal treatment under the law.

This is the same type of discrimination that is vilified by contemporary media. This has the same seriousness of discrimination as that based on genetic heritage or on belief.

The absolutely MOST discriminatory taxes (it's a toss up as to which is worse) are inflation (remember, inflation is set by government monetary policy), and the "Progressive Tax", or "ability to pay" system. In this system the most Productive individuals, as a class group, pay an inordinate amount of both money, AND a larger portion of their work life which is taken "for the public good". For a period of 5 or 6 months, all of their money is taken away. They must buy all of their things, their home, insurance(s), food, transportation, and so on, from only 6 (or less) months of their annual pay. Fully half of their life has been taken away. Every year, half of the efforts of all their labors are taken. Across their entire career, half of their entire life's work has been taken "for the good of everyone".

In the present US system, more than 48% of our working people pay no Federal income taxes. In reality, they are benefiting from the collective things, from our Military, from our public works, from our technological advances, and for local public education, without paying as much as one single day's work. In fact in the US, an un-earned bonus is given to each low-income individual. In fact and in deed a reward for their low Productivity. This gives a strong incentive to not contribute Productively to "the public good." "Well gee, if I work then 'they' take away part of what 'they' are giving to me now!"

WHAT IS FAIR?

The real issue here is "fairness". What is fair? Our progressive tax system is discriminatory. It's discriminatory against those who are providing more goods for a better standard of living for everyone. As much as some would try to convince you otherwise, success in business is not by luck.

Success in one's life work is not by luck. Success is achieved through motivation, learning, understanding, willingness, and in a few rare cases by birthright or through "connections".

The Federal Income Tax Rate for Personal Income for 2009 states that all income over $186,475.00 will be taxed at a rate of 35%. This is just Federal Taxes. That's $35 thousand for every $100 thousand earned. That's 35 cents out of every dollar. That's over one third of all income over the income mentioned. Out of a work year of 2,000 hours, that's a personal contribution of 700 hours, that's over 4 months of a person's life for every year! It's said that a thief steals more than just money from a person. They are stealing that same part of a person's life. They are stealing a portion of that person's plans for retirement. They are stealing the dreams of that person. They are stealing the dignity of that person and their right to ethical pursuit of happiness. Our country does not guarantee happiness, of course, but defends our right to pursue it.

So, the fairness and morality of a tax system has a profound effect on the future structure of a society, the total amount of the combined Productivity of the individuals in a society, the average standard of living of all individuals in that society, and the sense of Productive contribution to the society for each of its members.

**The bottom line regarding taxes is that every time we have Reduced the percentage of income taxes, absolutely every time, the actual amount of money collected by the government in taxes has skyrocketed due to the expansion of our Economy!**

Money circulates through society. The more the money circulates, the more money people are making, and therefore the more they are buying, and the more products which must be made, and therefore the more actual tax money is paid to the government. When tax percentages are increased, it acts as a damper to the circulating money, and taxes paid jump for a very short time, then plummet along with the Economy in general.

So in the absence of a Government itself being Productive and creating goods to support itself, some small portion of the Economy must be collected in Taxes to support our few, necessarily collective needs. Many say this is from 8% to 10%.

Replacing the current income taxes for businesses and individuals with a "Value Added Tax" of no more than 8% (fixed by law to not exceed 8%), would yield more tax income, would be more fair, and would still bring in more absolute dollars in taxes from the Rich, as the Rich would continue to buy more things. The existing income tax system for businesses and individuals must be totally abandoned for this to work. Of course States could choose independently to follow the same course, or maintain their current income tax systems.

States should never succumb to being dependent on the Federal Government for their revenues. Our States are, after all, our governing system, within a supposedly mild and simple framework of Federal oversight. Unfortunately, the Federal government, through mandates, (frequently unfunded), has continuously and consistently forced its agenda down the throats of the states.

Here's a simple example of the conservative "center of the road" philosophy, as contrasted to the liberal left.

You have a small "victory garden" in your back yard. You grow a few tomatoes, potatoes, beans, some peppers and cucumbers. Those fruits of your labors are solely the results of your efforts and GOD's grace.

1.) You have the absolute right to consume all the vegetables.

2.) You have the absolute right to defend your food from theft.

3.) No one has the right to take any of your vegetables, not even one single percent.

4.) You have the right to give some or all of your vegetables to anyone you choose.

5.) You have the right to sell any or all of your vegetables to whomever you want, at any price mutually agreed upon.

6.) If someone comes up to you and throws himself or herself on your mercy because they are hungry, you have the right to give or withhold your vegetables from them. Of course this shows the true meaning of charity, of the ability to decide for one's self to be charitable or not. If someone decides for you that you must give food, then they have taken away your right to give. They have taken away your opportunity to extend charity. Then, it's just another tax.

Can't we say the same for our earnings? Just because our efforts are through someone else or through a corporate citizen, should our money simply be taken and given to someone else? Of course not.

This is what Robin Hood was the champion of reversing. Robin Hood was against too much taxation. Robin Hood took back from the Sheriff of Nottingham to return the earnings to the people. Robin Hood won the admiration of Maid Marion for his good deeds as well.

"A wise and frugal government, which shall restrain men from injuring one another, which shall leave them otherwise free to regulate their own pursuits of industry and improvement, and shall not take from the mouth of labor the bread it has earned. This is the sum of good government." - Thomas Jefferson

This is what the American Patriots overcame in our Colonial times. Shedding the oppressive rule by an unjust and domineering government of England. This is the History of America. This is the core of our keys to success and excellence. This is what we, as Americans, are entitled to be proud of. This is nothing needing of an apology in the eyes of the World, but rather sets the standard for World Admiration and Excellence. It is "The American Dream" of limitless bounds, and not just this in theory.

## INDUSTRY & PRODUCTIVITY

Simply speaking, Industry is Productivity. Herein is specifically excluded the common use of the term "Industry" to include "service industry", "entertainment/sports industry", "financial industry", "transportation industry", and other support groups, some of which are of course vitally necessary for true Industry. Included in Industry are individuals and groups involved in *Producing* food, wood products, housing and furnishings, vehicles, energy, clothing, hunting supplies, electronics, appliances, tools, and countless other things.

Those individuals and businesses who Make "durable" and "non-durable" goods are Producers. A fine way to determine if something is a "good" is whether it can be sold after it is bought. You can sell a dozen eggs after you buy them, but you cannot resell a ride in a taxicab.

### ON ISLANDS FAR AWAY--TWO FAIRY TALES

If, on a small island nation, there were 100 people who were all prolific in their Productivity, there would be a lot of goods, nice houses, food, leisure and a wonderful way of life for everyone, regardless of the type of money system. In addition, the actual taxes to support the one government employee, would be an extremely small fraction of the total of goods produced. In this example, total taxes would be perhaps 1% or a little more.

On the same island with 100 people, if 99 were Productively engaged, there would be some repairs the Productive individuals couldn't themselves make, there would be a need for spiritual focus which the 1 government individual could not morally supply, and there would be someone for medical care needs beyond each individual's capabilities. So, in this example this would leave 97 people who were fishing, baking, supplying the goods, building buildings, and so on. And maybe there would be 10 or 15 retired people who had already made their Productive contributions during their working lives. There would still be plenty for everyone, and everyone could enjoy a good, rich life.

If, on another island, only 5 or 10 of the 100 individuals were productive, with negative incentives for others to produce, those 90 others would quickly lose the idea of, the feel for, and the joys of Productivity. They would lose sight of the pride a person gains through his or her Productivity, and as result there wouldn't be much of any goods for anyone except the few productive individuals. If everything were made collective and shared equally, there wouldn't be much of anything for anyone. Maybe the 90 others could provide some service like back-rubs or psychic services, but there would be little food, little housing, little fuel and little of anything else for anyone. This Economy would have no dependence on monetary policies or type of money, even if everyone were paid on a credits system, or a system using dried fish bones. If money were used, it would not matter the size of the individual units, the amount of inflation, or any other aspect of the money. There would still be only a limited amount of goods to go around.

If the combined Productivity of all individuals exceeds the needs of their society, they can then trade things for products from other communities. They can enjoy everything they need that they can supply for themselves, and in addition, nice other things such as spices or nylons, or efficient boat engines they can't make for themselves.

GOVERNMENTAL STIFLING OF INDUSTRY

In our Economy, we need to reduce taxes on Industry, which naturally creates jobs, which naturally takes people off of "Unemployment Insurance" payments, which naturally allows more people to earn more money and pay more taxes! No, it's not the same as increasing taxes on those who are working and increasing taxes on Industry. EVERY TIME WE HAVE REDUCED BUSINESS AND PERSONAL TAX RATES, THE ECONOMY TAKES OFF LIKE A ROCKET AND GREATER TOTAL TAXES ARE PAID TO GOVERNMENT. THIS HAS HAPPENED ABSOLUTELY EVERY TIME. Go ahead and look this up for yourself. More real need for people to work creates more jobs naturally, and more things (food, houses, technology items, clothes, everything) are made, and the Country is richer and more people can have more things and better medical care is afforded.

Success stories are never resultant of forced central organization of Production, nor motivation solely by "one's duty to the party", nor for "the common good". Central planning can only ever design and enforce standard rules. Exceptions to standard rules are normal for Free Enterprise Industry, and the key to America's Past, and hopefully future, Prosperity. Operating under standard rules only ever yields a standard, pathetic result. We see this in a casual glance at traditional Communism. We see this in a brief overview of West Germany versus East Germany. West Germany continued the technical, medical, and economic advances typical of the German People, and East Germany failed miserably under Communism. In

China, during Mao Zedong's Social Revolution, tens of millions of Chinese literally starved to death. China's recent move toward Capitalism and ensuing stellar performance is proof positive of the fantastic benefits of Incentive for Productivity.

This is a symptom of so many failed (but of course optimistic) designs for ultimate government control, and social design. These are the failed empires, which ALL ultimately succumb to revolt by the downtrodden masses. ALL of the success stories throughout history are associated with cooperative Productive ambitions, with Incentives proportional to Productivity. This is seen today in China, with many Productive individuals becoming rich, not just working themselves to death for society, (which is a normal way of life in Communistic countries).

How do we "Fix" the Economy? We don't "Fix" the Economy; we merely need to remove the restraints. We DO NOT need any further restraints. We do not need to cripple Industry with "Card Check", with "Cap and Trade", we already have many other restraints and types of oppression. We can't artificially "stimulate" the Economy by injecting effectively counterfeited money into areas with political "pork" to social groups such as "Labor", "Community Organization", and other "inside" entities. Creating 1,000,000 new government jobs will do nothing to increase the total of goods produced. It will merely take more from those who are Productive.

If we add 10 million "government jobs", absolutely nothing worthwhile is accomplished. It's the same as putting a little heater under a thermometer. The thermometer shows a higher temperature, but the room itself is no warmer.

ORGANIZED LABOR

When Industry (broadly all groups making things) have strong needs for people to work, Unemployment is low, and then it's a "Sellers' Market" from a potential Employee's standpoint, and better "conditions" and better "benefits" are needed to entice people to prefer one employer over another. This is the ideal situation, and it nourishes a successful society.

Today, and over the past 20 or 30 years, self-imposed "cost control" has been crucial for survival of Productive Organizations. Isn't this then (the federal preference for Organized Labor and forcing workers' rights), just another measure of the choking restraint of Industry? When Industry and Productivity fail, NOTHING, not even plundering ALL personal wealth, can prevent a national failure. So, we should not place artificial restraint and costs on Industry and Productivity.

People should work for a company. The company should be responsible for their employees. A Union is merely a Political Machine, which should not control a company, and should not control the supply of labor. OSHA, Minimum Wage, Child Labor Laws, Company's Apprentice Programs, and a multitude of sensible Government regulations have pretty much alleviated the need for Unions. Now, Unions have predominantly lost their Mission, and are, due to fear of obsolescence, of necessity becoming a Political Party in National Politics. Unions are, in fact, monopolies, otherwise banned in industry.

There are many who believe that the conflicting missions in "The Democratic Party's", support of those who don't work, and support for Unions whose people DO work, could cause the unraveling of the Democratic Party as the Unions peel away to start their own Party, or to become closer to Conservatives, who do value good work and good workers. It is of course the Unions' goal to become rich to increase their own managers' incomes, and to stay in business for the long haul, yet it pursues this goal under the guise of making a better life for their working members.

As US made goods become less and less competitive with those imported from other countries, it is difficult for Unions to rationalize to their members what they are doing as their members lose their jobs. Is it any wonder that our Big Industry builds new facilities in "Right To Work" states? To their credit, Big Industry recognizes the undue influence of Unions, and treats their employees fairly, so that their employees will not feel a need to Unionize. Joining a Union after all does require a moderately expensive monthly payment for workers. Monthly Union Dues range from about 1 hour of pay per month (0.6%) to about 3 hours of pay per month (1.7%), plus "fees and assessments".

In the early 20th Century, Unions played an important role, as previously mentioned, promoting the working conditions for many people, both Union and Non-Union. It's a pity that the senseless restraints Unions now place on Big Businesses tend to cripple some of them "from within".

A classic example is a "Fireman" on a train. When I was a kid in the 60's, all freight and passenger trains had already made the switch to diesel locomotives. Unions were still quite strong. The rules or "contract" (Collective Bargaining Agreement) between the Company and the Union specified that there must be a "Fireman" on all trains. I thought "what a good thing for safety". If a train should catch on fire, there was an expert on hand. And then to my utter surprise I found out that a "Fireman" on a train is the guy who shovels the coal into the boiler! What a nonsensical thing to do! Pay a Fireman to ride a train, when there was no longer a boiler! I met a guy, Paul, who had been paid for years to ride a train, and didn't really do much of anything Productive. Other rules limit how productive a person can be, in order to try to increase the size of the Union. The Union should rather be more worried about the survival of The Company, for existing workers to have jobs in the future.

Another example is the US Auto Industry. Long seen throughout the World as a standard of Excellence and Productivity, we had a little hiccup in our Economy, and two of the "Big Three" went belly up. With today's Government Leadership, the Unions have made out in a big way, with the Union itself now owning a large portion of General Motors. Seems a clear case of "the shoe now being on the other foot", and it should be an interesting time for them.

But, enough about Unions. We needed Unions at one time, and they brought about good change, but now...

PRODUCE OR PERISH

Let's say that Production in America totally stopped. Everyone involved in Producing in Industry, Farming, Fishing, Construction, Design, Creativity, etc, decided they had had enough of having their money taken and given to non-producers, and decided to stop work, and to get onto the "band wagon" of Welfare, Section 8 housing, Food Stamps, "WIC", Medicaid, ADC, SSI, un-earned payouts from Social Security, and so on and so on. It actually doesn't matter if ALL of the Service people continue to work, transporting empty cargo, treating patients, governing and administrating, cutting hair, practicing law, mowing grass, giving speeding tickets and arresting people. The courts stay busy, people talk on their cell phones, they go bowling, they ride their bicycles, go hiking, etc.

Then of course, we would need to import all goods from other countries. The "Trade Deficit" would become 100%. We would import absolutely everything. How long would the "Total Net Worth" last? Maybe a few years? Five years tops. This scenario ignores the Charity which people in many other countries would want to give back to Americans, that is from Europe, the Pacific Rim, Asia, Mid East, Ethiopia, Pakistan,

Somalia, and so on. (I wouldn't hold my breath.) So even though 85% of Americans involved in "non-Producing" might still be working as hard as they could, there would quickly become Zero net worth in the U.S. We would all starve to death, no matter how healthy we were, as we would have no money to import food or anything else.

Of course this is a ridiculous scenario, as people would start to grow their own food, cut down trees to make houses, go fishing, grow hay and corn for farm animals, make soap out of animal fats, spin wool into yarn for clothes, make buttons out of bones, make bows and arrows and clubs to protect their goods from theft, and become self sufficient. People who knew the value of Productivity would survive.

This is the story of American Native "Indians", the first immigrants and the original settlers of the North American Continent roughly 14,000 years ago. There was very little need for National Defense, and the political and legal systems only consumed a tiny fraction of a percent of the value of things produced. There were few doctors, no insurance salesmen, no stock market analysts, people hiked or rode on animals, and virtually everyone enjoyed almost 100% of the "fruits of their labors". People did not have children whom they could not afford to raise themselves. People who did not work were outcasts, and vagabonds, although some survived on the pity and charity of Productive People. Families taught their children the value, rewards, and necessity of hard work. People were free to "Pursue Happiness". No one was guaranteed happiness, medical care or a "life of leisure". People earned a living and saved for their old age, and passed along everything they saved to their offspring when they died.

Lawyers and "the state" did not get anything when someone died. Only a casket maker and perhaps a doctor earned a little from a person dying. In their "retirement," old people worked as much as they could, and lived off the savings from their previous Production. There were many terms used to ridicule those who did not want to work, and to those who had babies they could not afford. Families who lost their "breadwinner" were taken care of through charity, and some money was made through work the remaining spouse could perform.

Productivity and Industry and the total quantity of things made directly affect the average level of a society's standard of living. Think about the village of 100 people. If they make a lot of stuff, then there's a lot of stuff for everyone. IF VERY LITTLE

STUFF IS MADE, IT MATTERS NOT WHAT EVERYONE ELSE DOES. THERE'S STILL ONLY A LITTLE BIT OF STUFF FOR EVERYONE. There's no pile of "money" somewhere to spread around. Even if there is a money printing person, with large denomination printing plates, it's really irrelevant. There's still no more stuff than that which has been made. So, when we make more stuff, then everyone gets more stuff. There's more food, more medicine, and value generated for entertainment and other non-necessities.

It cannot be stressed enough how important Making Things and Stuff, "Production" - fishing, farming, mining, factories, and so on, is to our Society. Productivity is the only method of creating true value.

**The success and average standard of living of individuals in a nation is dependent only upon the total amount of everything that everyone makes in that nation, and the number of people in that nation who consume those things.**

If all people are involved in Services, then there are no Goods, and there is no actual value created to pay for Services.

Through increasing Productivity, more people become more satisfied with a better standard of living, and there is less discord in Society. There is less perceived need for Lawyers. Industry and all service segments of society are more prosperous. Banks do better. Repairs are better afforded. People buy more new cars, more TV's. "Wall Street" does better. More people can afford to go to Doctors. More people go to restaurants. More people spend more money on more things.

BIG INDUSTRY, SMALL BUSINESS

We are now getting a strong message from our media and our government that "Small Business" success is key to improving the jobs situation. Of course there will always be a need for small business in the Construction Industry, and Food and Service Industries. Are the Small Businesses the ones who made this a Great Country?

Was it the fishermen and trappers, the cobblers and candle makers, the family farms and salesmen who built this country into the global standard of excellence? In a word, No. It was the fantastic result of visionaries whose mass production processes turned our country from a Village/Trades Oriented small enterprise system into THE Exporting Giant of the World. It was our Industries, and here I mean Big Industry. It was those heroic producers, who single handedly fostered our specialization in mass production of energy, metals, machines, and medicines. It was their ability to make more things available, less expensively, than is possible through "small business".

Ethical companies, whose operations are based on a preponderance of real science, have constantly improved their processes and materials, and responded to the greater needs of society. Their fantastic Productivity must be recognized as the foundation of advancement of our society. In fact, Industry and Productivity as a whole are solely responsible for a society's standard of living. Everything else is in support of Industry, including Government, Finance, Labor, and, as rewards, the "entertainment", "dining", convenience, laundry, and other service industries. This is very easy to recognize when we take but a cursory overview of the conditions of people in various countries. When there are too many people and not enough Production, then there's lots of poverty and starvation.

It is Big Industry that has created a middle class. - A middle class of people lifted up out from poverty through Opportunity. Opportunity that Industry has provided for good, reliable, Productive work. The massive wealth that Big Industry has provided for Americans cannot be overestimated. Big Industry must once again become the "machine", not controlled by Government, not with artificial restraints as are applied by backward countries under Communism and Socialism - Not Government owned industries, not Industries suffering a chokehold of regulations and taxation, not Industries subject to massive employee costs such as those through Unions and unrealistic dreams such as "Obamacare".

It must also be said that it is Big Industry that enabled the North to win the Civil War. It was Big Industry that empowered the U.S. to prevail in World War II. Where would we stand if such a global calamity would arise today? Would we import our materiel from China and our oil from the Middle East?

INFRASTRUCTURE

It has been said that the majority of the "Infrastructure" in the US was built from the 1930's through the 1960's. Contemporarily, we find that we have many problems and failures of the basic installations of Electrical Power, Water/Sewerage, and Transportation Systems. Of course we must "fix" these things. The underlying question is WHY? Why has this system of Infrastructure become unreliable? Why has maintenance fallen behind?

We only need look at the state of industry during those times (1930's - 1960's), and the actual wealth we were generating. The percentage of people involved in Productive enterprise was very high. Industry (farming, mining, manufacturing, fishing, etc.) was THE employer. Industry and the goods produced needed a few people involved in information, distribution, and sales, and the people employed in Industry needed only a reasonable degree of support from services (including government). Most things bought were made in the USA. We had protective legislation that discouraged low cost, and frequently inferior, goods from overseas. We produced a lot of goods, and very little of the wealth generated was sent outside of the US. Production generates a Nation's wealth. Reducing Production by whatever means, whether intentionally (perhaps sinister - vilification of Energy, for instance), or by act of nature or by deliberate competition by availability of 3rd World Goods, causes the Nation's Wealth to suffer.

We had a lot of wealth, and we could afford a high degree of infrastructure investment as a society, paid for by the Productivity of Industry. We had the Space Program, beautiful new highways, explosive growth of "The Grid", systems of clean healthy water, a system of Public and National Parks without entrance fees, and we set the Global standard for "average standard of living", in most cases with only 1 person per household in the workforce. At this time, in the 50's and 60's, roughly 4/5 of our workforce was involved in Production in Factories, on Farms, Fishing, Transportation, Energy, and so on.

Then, something happened. Today, only about 1/5 of workers are involved in Industry and Production, a stark contrast to the highly productive era. Today, a significant percentage of the population is paid to not work. They are paid to have babies, they are given free food and free healthcare, nearly free housing, and this has become too big a burden to bear. Who pays those costs? The government doesn't have any money itself, as it does not Produce anything, except for a few leases on resources. ALL wealth is generated by Industry and Production.

Tens of millions starved to death under Mao's rule, but in the post-Nixon era, China's Productivity and culture have blossomed. Of course some say China's values and implementation of their ideals is not perfect, but we must first look at the tremendous progress that has been made by "heading in the right direction", and actual improvements that have been made in enabling freedom and personal "pursuit of happiness". Recognition of the values of Productivity and personal profit motives have made all the difference.

Interestingly, the "American Worker" is the most productive in the world. The problem is that we don't have enough workers involved in Production.

Today, all Infrastructure budgets beg the question, "How do we reduce costs?" We don't have the money to pay for adequate maintenance. We delay the next inspection, we delay the next repair, and the system becomes riddled with holes and inadequacies. On top of that, our population has skyrocketed, and this has put a tremendous strain on our Infrastructure, which was never intended to handle the present stress of use.

So, we are not a wealthy country anymore. Our National Debt is itself climbing through 100% of our Annual Production. Should we institutionalize Industry? Can we dictate when and what our factories should make? Leaders do just that in Communistic and Socialistic countries. Does it work? Just look at the people. Look at N. Korea compared to S. Korea. Both countries have the same heritage, and the same basic natural resources. It's obvious that if we want a wealthy country we must encourage, not "regulate to death", our Industry. Recently, we have embarked on an incredible trend toward over-regulation, rationalized by "emergencies." Emergencies of healthcare, the economy, energy, banking, the environment, education, and whatever else Liberals feel they can control.

Our failing Infrastructure is only a symptom of the loss of wealth of the US. It's a symptom of the failings of society caused by the Socialistic diversion of wealth and the failure to recognize and allow the natural rewards to Productive people and Industry. Earned and retained dollars are a positive incentive to work harder. We need to allow the productive people to be more Productive, and to act as real, live models and mentors for others.

The bottom line is that we can no longer afford any but the most compelling expenses for Infrastructure. The bottom line is that our Production per American has shriveled to a prune of its former glory. The reality is that far too many Americans are employed in Service, in Government, and receiving Welfare in a broad spectrum of flavors, from a wide variety of Public Sources, of which there are too many to list here. Our Government is growing at an unsustainable rate, well in excess of the growth of our population.

This, then, is the reason for our "Failing" Infrastructure: Much less actual National success and therefore less tax is generated, and that which is, is directed elsewhere. So it's not "Rocket Science". You spend less on upkeep of your house, then it gets ratty. You spend fewer resources on your lawn, and it gets crappy.

We have yet another "Stimulus Bill" getting ready to pass Congress: an "Emergency" expenditure for Infrastructure, and at a time when we are having more Infrastructure failures. Why should we be suddenly having more Infrastructure failures anyway? Are they just a focal point for the media? Or are there really more failures?

Regarding where we are directing Infrastructure monies, we need to make sure that the money is not merely a payback to Unions or for other Political purposes. We're talking about $60 Billion to be spent. One of our biggest Infrastructure failures is in not protecting our Country against invasion by Terrorists and from Economic Invasion on our Southern Border. And we have just directed only 0.6 Billion in that direction. Most Americans believe money spent to exclude Economic refugees and Terrorists from entering illegally across our Southern Border is one of the BEST Infrastructure elements we should spend our money on. Why are we not spending 10's of Billions there? Savings to our entitlement programs, safety for Americans, and jobs for Americans would surely benefit by a 10:1 factor - Let's fix that Infrastructure problem First. Let's build an impermeable physical barrier.

CONCLUSION

A recent interesting creation, "The Story of Stuff" is being promoted by "The Joyce Foundation®". This cute cartoon implies, no states, that we must consume less manufactured goods "to save the world".

Goods, and particularly the Industrial Manufacture of Goods is the singular thing without which we would literally still be swinging naked through the trees, yelling for others to stay away from "our" fruit tree. Is this one of the goals of the Joyce Foundation®? To send us back to "the good old days"? Of course then again, if we consume less, then fewer goods will be manufactured here, and there will be even less for each person, and particularly irksome, there will be less for those who do Produce as theirs is taken for the non-producers. Of course at the present, productive and working people are a strong social force. We are too strong to be taken over by an unfriendly government. Is this the rationale of the Joyce Foundation®? You decide for yourself. There are many groups and too many "non-profit" groups who wish to change our proven way of life - Our system of freedoms and our system of Productivity and Rewards.

Somehow, over the past few decades, Big Industry has become perceived as the "enemy", the defiler of our environment, an example of corporate greed run amuck. This carefully planned scapegoating has provided ammunition for those who would further their own agendas for their own benefit.

Productivity should instead be not discouraged, not punished, and we should allow its various blessings be naturally sprinkled among us through Opportunities for each of us to contribute and be also Productive.

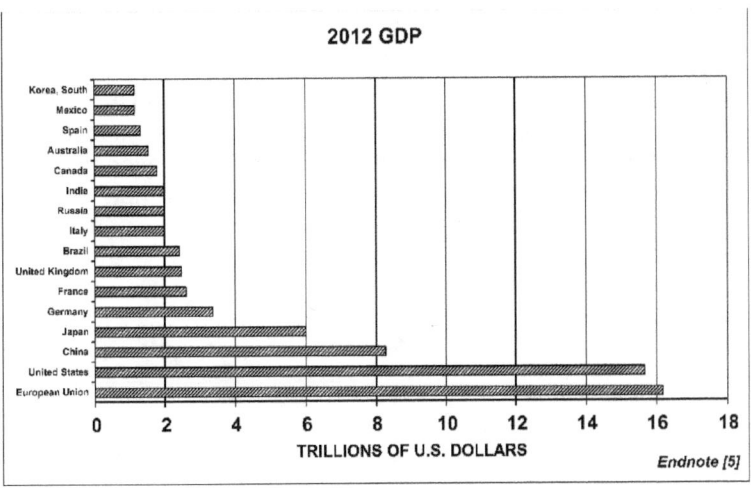

**2012 GDP**

TRILLIONS OF U.S. DOLLARS

*Endnote [5]*

America's success, measured by Gross Domestic Product (GDP) is, even in 2012, still the largest in the world.

This level of excellence has allowed America to extend charity, both by US Citizens and by our Government, to "The World" on a scale heretofore unimaginable. This amazing success Is due in major part to the fantastic productivity of large scale Industry, and in particular, Heavy Industries such as materials, energy, and transportation, and Large Industries such as medicine, information technologies, and food and drink. These companies have provided the bulk of the value which provides employment and which pays for the services industries and our abundant charity.

Forcing standardized rules and controls, disproportionate

taxation (50% instead of 10%, for instance), quashes the visions, dashes the hopes, squelches the Productivity, and provides less for ALL. Controllers are NEVER the ultimate providers. They can only hope to take from some to give to others in the name of some "glorious" cause.

Implicit in the view of the virtues of opportunity, hard work and Productivity are the beliefs that "more for everyone" and a higher average standard of living are desirable for a society. When most people are well off, have a broad range of their GOD-given freedoms and are comfortable, it's hard for any government to force a population in a different direction against their will. Decisions and Laws made against the "Will of the People", (what some would try to lead you to believe are merely "necessary tough decisions"), is in direct contrast with the spirit and intent of The Constitution of the United States and the foundation of our Great Country.

Incentive fires the imagination, and things which couldn't seemingly be done, are done. More things made yields more things for everyone. Every technological advance, every medical step forward, each improvement in efficiency, each new creation born of a vision for a better future creates a better condition for each individual and therefore for our whole country, humanity, nature, and our Spaceship Earth.

# THE MELT-DOWN & BAILOUT

The decreases in family income over the 2007-10 period were substantially smaller than the declines in both median and mean net worth; overall, median net worth fell 38.8 percent, and the mean fell 14.7 percent. *[6]*

What really happened to cause the 2009 Economic Meltdown? Why did this lead to a recession? What about those "excessive bonuses"? Who is to blame? Was this a carefully crafted plan?

Simply put:

1. Some lawmakers, including Jimmy Carter, in a "Progressive" move back in 1977, believed that all Americans should have a shot at "the American Dream": to own one's own home on a nice street with flowers and a white picket fence lah-di-da. Those "liberals" or "progressives", or whatever name you want to give to those who believed that lack of a family's Productivity or Resources shouldn't exclude them from home ownership, passed a seemingly innocuous bill: "The Community Reinvestment Act". This legislation dictated that lenders extend credit to people whom lenders had determined, through their tried and proven strategies, were at high risk of default. These extensions of credit to unqualified buyers posed a risk to the lenders' continuity of business. Financial Institutions were forced to lend to people who did not previously qualify for mortgages. This law was based on claims that lending institutions were discriminating against certain communities. This law directly coerced sensible business people to conduct part of their business in a way that was truly irrational, in an essentially charitable fashion. Incentives to operate their business this way, through "HUD" (Housing and Urban Development), and virtual guarantees by "Fannie Mae" and "Freddie Mac" convinced a number of lenders their risks would be limited. If Banks did not lend the money, the law had teeth in it that could make the Banks' life difficult. This is what caused the "Sub Prime" crisis. All of the subsequent actions, which many would try to convince "the masses" to believe were the cause of the problem, were in fact not. If banks had NOT lent money to folks who did not qualify, there would now be no problem. There would be no Melt Down, and no need for "Financial Reform". Financial institutions were first forced, then financially rewarded, for acting against common sense and the proven principles of their industry.

In considering a credit application, financial corporations consider not only a family's ability to hold up their end of a mortgage contract, to repay the loan, but also to assess the downside of that contract, to answer the question: "What can we (the bank) do if the homeowner defaults?" The loan is of course secured by the property for which the loan is made. So, if the family doesn't make payments on time, then the property is "foreclosed". The bank legally assumes ownership of the property, and sells it to recoup their money "without loss". For each foreclosure, however, the financial institution actually suffers a loss of $30,000 to $50,000. With the MASSIVE defaults, property values were much lower than the amount of money the Banks lent to buy them, so the Banks lost massively. **In trying to accomplish a social dream, the Government actually caused the financial meltdown.**

It is, after all, the business and the purpose of Banks to be able to provide a return on investment to their share-holders, pay their employees, pay for their expenses, pay their taxes, to hopefully expand their business, and to improve the general condition and wealth of our great country and citizens. Banks, although a service organization by definition, help enable the expansion of Productivity in our country. It is not the purpose of a bank to conduct its business in a way to lose money, and therefore to go bankrupt itself.

Financial institutions were "incentivized" to make loans against their better judgment. Extending mortgages in low-income areas, loans secured by property which could not reasonably be sold through foreclosure, which in a predominance of cases would be "bad business", was thereby mandated on Jimmy Carter's watch through this legislation.

"There are two ways to conquer and enslave a country. One is by sword, the other by debt." - John Adams

2. Perhaps US taxpayers would get some joy that this "forced charity" would "kick-start" those with a little "seed-money" into a self-improvement whirlwind; that the people in those communities would take over with joy and optimism, with a spirit of revival and improvement, and would boost themselves up to a level of self-sustainability. However, the plan was not successful. Our low-income communities did not self-start into paragons of model communities. The people in those areas did not imagine a brighter light on the horizon. They did not gain a sense of self-improvement, confidence and glory in their own cause. They did not embrace the virtues of the age-old moralities of education, hard work, and accomplishment. We of course can't ignore and must admire the cases of personal determination and success in

these areas, as we always have. As much as its proponents would have liked it to seem to, "The Community Reinvestment Act of 1977" just did not have much of an effect. Even now, the sad situation is that the peddlers of this myth believe they did well. The people who forced this legislation FAILED MISERABLY. These Legislators are directly and inexcusably responsible, and must be made aware of their Failures and their dis-service to our society. Sure, they thought they were doing the right thing, but so did Jim Jones and so many other insane ideologues.

3. ACORN® to the Rescue! Backed by proponents of "radical economic justice" via US taxpayers' funds, this group, which is without any doubt a politically aligned organization with a bent for Social Reengineering, used tactics of collusion and intimidation which were by definition terroristic. In addition, they used carefully concocted legal maneuvers to have their way with our society. Using questionable, politically aligned tactics to change the composition of our lawmaking bodies, ACORN® embarked on a vengeful track to force an exponential expansion of the Community Reinvestment Act. They did this with the complete support of legislators such as the (then) Senator Barrack Obama (as he stated on a 2007 video recently brought to public light). ACORN® was in 2009 proven to be working outside the legal/moral standards of our society and subsequently Congress was shamed into de-funding their operations with "government", that is, taxpaying individuals', money. As was predicted by a number of Conservative spokespersons, ACORN® is not disbanding its organization and mission, but is in fact morphing into less recognizable, smaller, regionally based operational units, barely under the surveillance of public view, but still funded by taxpayers.

So, through frightening techniques of invading boardrooms of banks and other financial institutions, through intimidation, harassment, and even stalking financial executives, ACORN® coerced the Banks to make even more loans later deemed "sub-prime loans". Banks, particularly large financial institutions, felt a strong need to comply, to avoid what was becoming to be seen as potentially bad publicity, but how could they possibly loan money with such a high likelihood of default and loss? How could professionals in ethical banking businesses possibly rationalize making such risky loans?

4. Fanny Mae and Freddie Mac join the rescue! Fannie and Freddie, for some unknown reason, (perhaps a sense of fulfilling a moral obligation or perhaps due to further political pressures)

agreed to buy up virtually any and all loans, including those with a 100% likelihood of default. So, suddenly, and as if by magic, making risky loans was OK, and there could even be profit in it.

If someone is sprinkling gold nuggets everywhere, and Fannie and Freddie say it's OK to collect all you want, guess what? That entrepreneurial spirit innate to Americans kicked into overdrive. Cell-phone salesmen became mortgage agent millionaires overnight. Financial institutions reaped in the big bucks. People moved into their magical new homes. Those people were of course excited and their feelings of skepticism soon vanished. They had "special elves" who would help them stay in their homes, which they had been convinced that they deserved.

Of course those twinkling feelings of magical home ownership soon begin to tarnish, as the monthly obligation continued to challenge their resources. Terms of "teaser rates" and the contractual agreement of future higher rates were ignored by people who felt that getting a start and a sense of future financial success would somehow make everything OK, are what directly led to the condition of massive foreclosures.

5. It's a lot like my friend Horatio tells me: "Those people on welfare didn't design the system, they are merely taking advantage of it." The Financial Institutions didn't design the sub-prime system; they just took advantage of it, against the judgment of eons of common sense in their business. BUT, Fannie and Freddie said "it is OK, and it is good", a few financial "wizards" were getting rich, and as a result many families were on the cusp of "The American Dream".

6. Then came the sudden increase in the cost of energy. America's failure to become energy independent, through use of our own vast energy resources, (as was the original mission of the Department of Energy), was the "straw that broke the camel's back". As the Election of the fall of 2008 loomed on the horizon, most Americans were not aware of the impending economic crisis, save for a few ineloquent economists, until certain Democratic Party leaders frightened us into the emotional state which lit the fuse of the "melt-down". Similar to the fear which led to the temporary blip in our economy caused by the religious extremist attacks of the September 11 travesty, Americans' feelings of security and growth were suddenly dashed, resulting in what was to become the recession of the decade. The condition created by the social warping of our biggest financial institutions built up a multi-trillion dollar

imbalance in our economy, which prevented our "can-do spirit" from allowing us to recover quickly.

7. So, the cat jumped (or was dumped) out of the bag. Prior to the November 2008 election, a number of people were screaming about how terrible "The Economy" was, and that it was going to fail, but it of course had not yet failed. What a perfect ending to a long stretch of economic good times. What a time for voters to think less about the virtues of Capitalism, to get concerned about the bottom falling out, and a time to put "Progressives" (whatever that means! ) in control. Their icon had been an image of motherly care, at the cost of those inconsiderate bastards with so much money. But the Economy hadn't yet failed! Where were those doomsayers when the Economic collapse was still in the womb? Were they depending on, or even planning this massive problem? Certainly we have heard many times Rohm Emanuel's quote: "You never want a serious crisis to go to waste, and what I mean by that is an opportunity to do things that you didn't think you could do before." Do what things? To push through a social agenda, which Americans do not want, and would during "normal times" never allow to happen?

Hummm. Economic Troubles = Liberals get elected. Hummm. More Economic Troubles = More Liberals get elected. If I were a Liberal Candidate or a controlling influence within a Liberal Party, I know what I would do to "improve" election results. On the long term, over periods of decades or longer, I would scheme to set up traps to discredit Conservative Principles and Free Enterprise. Particularly, I would attack "Big Business", which cannot easily be controlled, and promote "Small Business", which can.

IF we had no "Community Reinvestment Act"; or

IF Fannie Mae and Freddie Mac had not held out huge carrots for financial institutions to act against their better judgment; or

IF ACORN® and ACORN®'s lawyers had not sued and terrorized our banks to force them to give out more bad loans; or

IF Americans who could not afford to buy a home were not led to believe they could by ACORN® and others,

THEN, there would have been no Economic Turndown or Recession, and probably more Conservatives would have been elected to office in the November 2008 elections.

8. Several big and many small financial institutions began to realize how much their balance sheets truly reflected the risk of bad loans and magical "derivative financial instruments". Uncle Sam to the rescue. Loan big bucks to bail out the institutions, which are "Too Big To Fail". But, wait a minute, what about those financial institution CEO's who are making such huge "Bonuses"? How could anyone deserve bonuses during a time of failure of their institutions? Let's make the public think that it was those "greedy" CEO's and greedy "Wall Street" who caused the economic morass.

9. Back to 1993…Slick Willie was in charge. Executive salary was limited by designating any salary paid to any person over one million dollars as not a legitimate business expense. What could companies do to keep their most important personnel? Many of these companies had "performance bonuses", which rewarded their top people for excellent company profits in a year. But, how could these same companies keep their best people, to streamline their businesses for future good business, when perhaps they had fallen behind due to inexpensive imported goods, or changes in consumer preferences or raw material costs, or increases in the costs of personnel benefits? Many of these executives were essential and crucial to their company's survival, and therefore responsible for maintaining jobs for their faithful and Productive employees. So incentives other than salary were given to hold onto those top performers.

What was the reason for Slick Willie going along with the idea of limiting executive pay anyway? What was that all about? Surely a slap in the face of the ever-growing reputation of American as being a place of a life without limits, if a person worked hard enough, and was productive enough. Companies were desperate to keep their top people, so they came up with a "workaround" for the million-dollar cap. Just give their top people attendance bonuses in addition to a million-dollar salary.

Here's my solution to the "bonus", but in reality "bonus in name only", or "B.I.N.O" problem:

Get rid of the income cap for salaries. Putting artificial limits on the maximum a person can earn is in total opposition to our elemental values as a free market society. It makes one think about the really scary historical records of regressively oppressive "governments" in ugly days of times long past. "Performance bonuses" should continue to be incentives for excellent business performance.

This failing of our Economy was totally due to un-natural manipulations of a normal, healthy, free market system.

Ultimately, some who felt they were "doing the Lord's work", or helping to transform our society into something more "equitable", ultimately led to our downfall. Will they ever admit their errors and wrongdoing? Or, are they just screaming even louder that we haven't done it right yet? You make up your own mind, but I'm more than happy to go forward with what we know works, then gradually think out and evolve step by step to a better society, one with more things for more people, with more and more people involved in Productive enterprise, and less and less involved in the non-productive.

So, the Economic Meltdown and "Wall Street Bailout" were believed by some economic experts to be initiated by actions of Liberals, with traps, legal maneuvering, monetary incentives, and mob-type intimidation of reasonably ethical Financial Corporations by Liberals and their agents. What other schemes are now in the works?

WALL STREET

There was no "Wall Street Bailout". This term leads most to believe that our superior Capitalistic System failed, and the American People had to make emergency loans to the entire American Economic System. This is an absolute distortion of what happened, and those who actively promote the idea of a "Wall Street Bailout" are in fact betraying the trust of The American People.

"Wall Street" generally refers to the New York Stock Exchange, which is located on Wall Street, SoHo, New York, NY and also to large companies who have offices literally on Wall Street, which is about 8 blocks long.

As an interesting aside, in Colonial Times, Manhattan Island was occupied by but a few farmers, who raised crops and agricultural animals. Some animals needed to be fed, but others roamed freely, restrained by the water all around the island. Pigs and possibly others were given free reign, eating insects, nuts and roots. When someone needed a pig to eat, they just corralled one up against a natural barrier, then killed and butchered it on the spot. It turned out that the swine were particularly destructive to the crops, so a fence-wall was built to keep the pigs out of the crops. A path, then a trail and a roadway and eventually a street followed the fence wall. In fact, that street is today known as "Wall Street". Manhattan was surely once known as "Pig Island" (but who would ever be able to prove that!), Manhattan and especially Wall Street have evolved to be one of the major centers of trading in our country and the world.

A few banks, financial institutions, and 2 car companies received emergency loans. Most of the loans, plus interest, have been repaid. The "bailout" was not a block of grants, but rather emergency loans to these companies. Most of the companies gladly accepted the loans, but in the American Dialogue, significant evidence indicates several companies did not want the loans, but were forced to take them! What was this about, and who stood to gain from this? What perception could be generated by more companies accepting loans? Would this be part of an attempt to "prove" that more strict regulation is required?

For the sake of argument, assume there were eleven companies that received emergency loans. On the New York Stock Exchange, over 8,000 different issues (company stocks, bonds, etc.) are traded, bought and sold. That's one company that received an emergency loan for every 700 that DID NOT receive any emergency "bailout" loan.

So one company in every 700 listed on the New York Stock Exchange received an emergency loan, and all of a sudden it's a "Wall Street Bailout"? That is totally wrong and distorted. It's obvious that anyone who believes that Wall Street was bailed out is totally ignorant about the business world, and the number of companies that actually received a loan. Or, those screaming "Wall Street Bailout" the loudest KNOW it's wrong, and want to try to make all Americans victims, to blame Capitalism. This is also wrong, and is close to being evil. No, on second thought, it is evil.

Let's look at the furor orchestrated around AIG. The US Government lent 170 Billion dollars in emergency loans to AIG. And yes, that's a lot of money. Bonuses for top executives were $163 Million. Yes, 163 Million dollars is a lot of money for any few people. Let's look at media focus and what our administration is trying to convince "the people" is a big deal. These "bonuses" were 0.09% of the total amount of the bailout loans. For every 10 dollars of the emergency loans, less than ONE CENT, that's $ 0.0096 dollars, was paid in bonus. I believe it makes more sense to concentrate on the remaining $ 169,837,000,000.00 of the emergency loans. Let's focus our attentions on the $ 99.90 out of each $ 100.00 dollars of the loans, not on the ten cents.

This is manipulation by jealousy and rage, to make a villain, to re-direct attentions, not a real issue about an "unintended" loss of a microscopic portion of taxpayer funds. We have sayings about "not being to see the forest for the trees", and "missing the big picture", but why are we being "told" to do this? To focus our attentions on minutiae rather than the important things? Some say it is a way of distraction, to slip through some other agenda item, un-noticed, while most people have been emotionally whipped up about a trivial item. It's easy to scare people, and very difficult to unscare them. The biggest driver is likely an indignation that no person deserves more than people in Congress can earn. Talk about a crazy arrangement! The people of the US, the bosses of Congress, do not determine Congressional salaries, Congress chooses its own salaries, increases, and expense accounts!

## BIG GOVERNMENT, BIG MONEY, BIG POLITICS, BIG BROTHER, BIG RELIGION

### RIGHTFUL / WRONGFUL ROLE OF POLITICS & GOVERNMENT

"Politics" is an act of public (human and many other animals) decision influencing processes and activities toward gaining support from others for issues, causes, and sometimes goods. To sway or convince a group of people toward favoring an ideal or a goal is "Politics". Throughout human history, we cannot categorically say that results gained through "Politics" are either good or bad, unless we see them either favorably or unfavorably as part our own personal view. Are "Liberalism", "Capitalism", "Socialism", "Totalitarianism", "Communism", "Religionism", or any of the other "-isms" good or bad? People have divergent views on these things. We can look at them and make up our own minds individually, or we can blindly accept what others say that our opinions should be.

Which do you think is better? Do you want to make up your own mind, or do you want to avoid that responsibility and give up that right to someone else? This seems to be the gist of "Politics". Politics is often more a game of immediacy, psychological content and manipulation than is involved in the long-term, rational unchanging "ground rules" and morals of a "good" society. - Rules such as "everyone pulls their own weight", or respecting others and their property, etc. Today, we seem to have a classic, virtually poetic struggle between acting on "emotions" versus "common sense". As said elsewhere, decisions based on "emotions" are immediate reactions, and "common sense" or "rational thought" takes time. So, promoting a "emergency" can become a tactic. Case in point, "Universal Health Care", which has been pushed for more than 50 years, is all of a sudden an emergency. An emergency only to the proponents, who realize that "common sense" has dominated the issue for more than 50 years. The sweeping Universal Health Care Bill was shoved through in opposition to the will of the majority of Americans. Most Americans did not want this law.

"I predict future happiness for Americans if they can prevent the government from wasting the labors of the people under the pretense of taking care of them." - Thomas Jefferson

After all, the individuals ARE the nation, and they need but a very small group of people in actual government to help maintain standards for commerce, to provide for common defense, to study and create knowledge to protect against proliferation of disease, and to act in our citizens' best interests, per our direction.

A LOOK BACK

Let's look at what our Federal taxes should actually buy for us:

National Defense

A National monetary System

A system of Protection of Individual rights

A national Postal system

Where did this list originate? The US Constitution. The Constitution grants ALL other areas of societal rights and responsibilities to our citizens and their respective Local and State governments. Many of our Founders felt the Bill of Rights was unnecessary, since if a right or responsibility was not on the above list, the Federal Government, by the language of the Constitution, was barred from getting involved in it. Other Founders felt a second emphasis was necessary, and added the Bill of Rights so that there could be NO QUESTION on the limitation of Federal power. The Founders knew that decentralized governance was not only less intimidating than a "Federal" structure, they also knew that decentralized governance was far more efficient. They felt that local government would be more in touch with the will of the citizenry and would also be more easily reigned in, when necessary. Our Founders were terrified of a Federal Government run rampant. - Rightfully so! The last of the original Amendments of the Constitution (The Bill of Rights) states:

*"The powers not delegated to the United States by the Constitution, nor prohibited by it to the states, are reserved to the states, respectively, or to the people."*

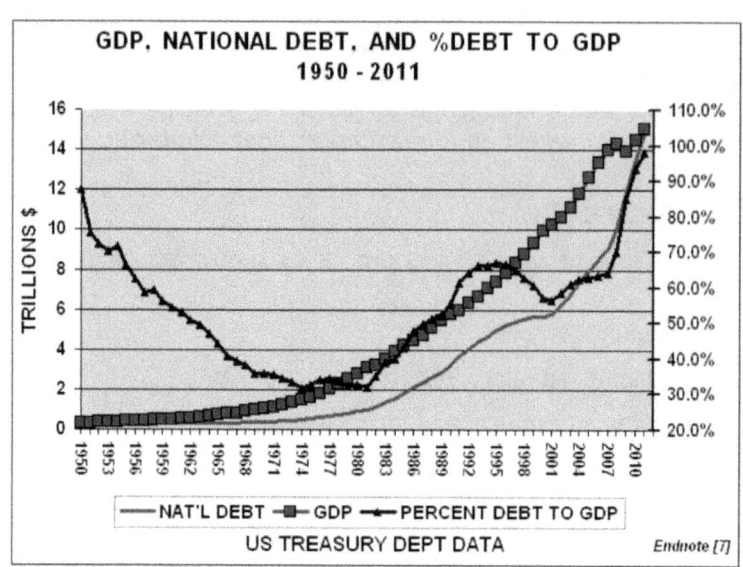

GDP, NATIONAL DEBT, AND %DEBT TO GDP 1950 - 2011

NAT'L DEBT — GDP — PERCENT DEBT TO GDP

US TREASURY DEPT DATA

*Endnote [7]*

The Federal Government has widened its scope of governance over our lives since 1789. Some of the additions (and rare subtractions) to their power are perhaps justified. The expansions which seem justified are those dealing with Anti-Trust, Transportation safety (in those industries that provide National and International services), International contracts and treaties, Interstate commerce and infrastructure, immigration, and to some degree, Public Safety.

"The Constitution is not an instrument for the government to restrain the people, it is an instrument for the people to restrain the government - lest it come to dominate our lives and interests." - Patrick Henry

Cabinet Departments: Year Established

State 1789

Treasury 1789

Justice 1789

Defense* 1789

Interior 1849

Agriculture 1889

Commerce 1913

Labor 1913

Health and Human Services 1953

Housing and Urban Development 1965

Transportation 1966

Energy 1967

Education 1979

Veterans Affairs 1987

Environmental Protection Agency † 1990

Homeland Security 2002

[8]

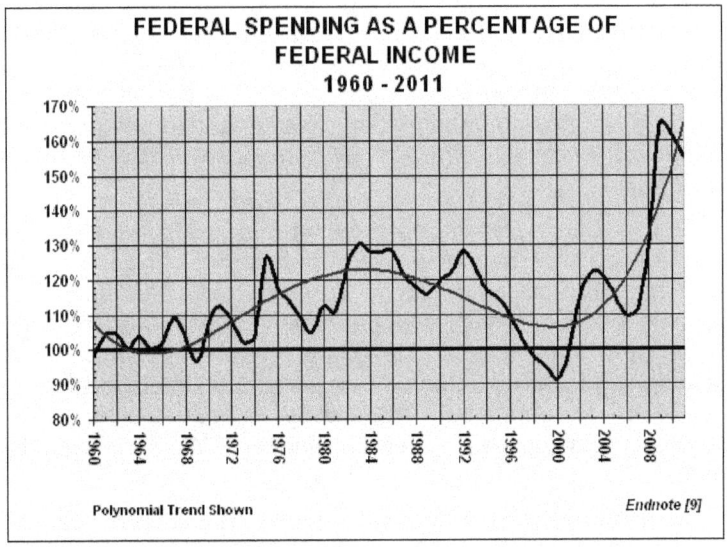

**FEDERAL SPENDING AS A PERCENTAGE OF FEDERAL INCOME**
**1960 - 2011**

Polynomial Trend Shown                                    Endnote [9]

It appears our Federal government is spending $1.60 for every $1.00 "earned". How long can this possibly last? How well would this work in your household or business? Perhaps they're unaware of the situation, since our Congress has been unable to pass a Budget in years.

"We must not let our rulers load us with perpetual debt. We must make our selection between economy and liberty, or profusion and servitude." - Thomas Jefferson

It is Government's job to protect and to gently steer our Economy away from bad practices such as Monopolies, nonsensical distribution of credit, and to plan against toxic pollution ($CO_2$ is NOT pollution).

We also must be protected from invasion by foreigners. Immigrants coming in must be controlled in rational numbers, and they should be allowed immigration for a sensible purpose, and for a specified period.

BIG MONEY

Our elected officials are responsible for attending to the business of Americans. We have chosen these people, and pay them to do their job, the job of and for THE PEOPLE. We should not pay our elected officials to do things for people in other countries. We should not pay for vacations for our elected officials. They should pay for their own vacations, as do you and I. We should not pay our elected officials to do work for "non-profit" or charitable organizations. We should not pay our officials to leave their station and go off in support of the DNC or RNC or the election campaign of a candidate. We pay our officials to go about the business of creating, administering, and defending our system of laws as directed by American Citizens & the Constitution.

Our elected officials are responsible for conducting our important work of maintaining the operation of our Locality, our State, and our Country. It is NOT the work of our officials to work part-time on our behalf, and part-time for their political party. Some officials are seen to place greater importance on campaigning for candidates of their party, than in eliminating the roadblocks to improving Productivity of our Country. This practice is totally disrespectful of Americans. We must stop this practice.

We must be careful to avoid improper influence by "Special Interest Groups" such as labor unions or financial groups. These groups often have their own "representatives" approach various Government Officials for some consideration in legislation. Some of our elected officials have also been directly influenced by money or guilt, or by other means.

The biggest singular external influence on our Government today is not the national league of lawyers, although its members do compose a large portion of our legislators and other officials. The biggest special interest group having excessive influence on our legislative process is not energy companies, the Better Business Bureau®, or Farmers or Industry groups, or even Environmentalists. It is not the Christian Coalition or Jewish League. It is not even the mysterious and powerful secret societies, or fraternal organizations such as The Masons. The biggest outside influence groups are hardly seen, and are now "Hidden in Plain View". Those groups are The Republican National Committee and The Democratic National Committee. How do they control legislation? By directing legislators who are controlled by their massive "seen" and more massively "not seen" finances. They are influenced and controlled by coercive semi-legal and illegal means. I can see it at work in my mind's eye. "Well Bob, you really did well last election with our contribution of $20 million and the help of many of our speech writers, our political consultants, our plants in "their" gatherings, and our investigations into the personal life of your opponent and their friends. Now Bob, you know we're having a real tough time with those Tea Party activists, and we're worried about passage of "Universal Health Care", the centerpiece of our plan for social justice.... So, I know we can count on your vote FOR the health care bill. Don't listen to those conspiratorial public opinion polls, they don't tell you how to vote, WE DO." Of course if this type of talk is not motivating enough, the pressure will surely become more intense and irresistible. .

"Resistance to tyrants is obedience to GOD." - Thomas Jefferson

Although many facets of "Politics" and the psychological coercive techniques inherent to politics are seemingly necessary for our government and our politicians, they are not legitimate functions of governing, and only act to circumvent the Freedoms of Americans.

Lincoln did NOT say "...A government of the Party, by the Party, for the Party." We vote for people, not for a Party.

## LAWYERS

What if most members of government are Christians in a predominantly non-Christian country? What if most members of Government were members of the BrickLayers Union? What if most in Government were members of the Malamute Trainers Guild? What if Kindergarten Teachers had a disproportionate presence in Government? What if Lawyers, who represent 0.83% of the American Work Force, represented 30% of our Congress? Our Congress is in fact comprised of 30% lawyers on average, with a larger share of representation in the Democratic Party. Perhaps legislation passed could have a tiny little bit of preference for the Guild of Lawyers and possibly enhance income for their members? This is after all, a ratio of 36 times their representative numbers in the American Work Force. Well, OK. Maybe Lawyers are more "qualified" to make laws, so let's allow double their percentage in "the public sector". Let's allow 1.66% of Congress to be Lawyers. Of course if 30% of the workforce were Lawyers, then that would be a different story. Maybe Lawyers could sue more Lawyers, and really increase our GDP. Then we'd be a rich nation once again. Or, is there something "a little fishy" about this logic? Figure it out. You've got a brain, use it! As an interesting side-note, the US has about 5% of the World's population, and some experts say 40% of the World's Lawyers. Does this make sense?

Many people are now questioning the proportion of "Career Politicians" in our government for these same reasons. They have superior incomes, superior benefits, and demand superior treatment. Congresswoman Pelosi, for instance, spends how much in public funds EVERY WEEK to commute from her home to her job? New Big Jet, non-stop, because she didn't like the inconvenience of stopping for re-fueling in the business jet she inherited with her job! What's wrong with her commuting home once a month on a public train? We'll pay for that, first class. Perhaps she can have a private car on a regular train schedule. This would be more than fair. If she needs to travel internationally, outside N. America, she can fly first class on a commercial flight. Plenty of security there. People who ascend to such "lofty heights" refuse to understand Real America. They'd rather make decisions favoring their personal benefit, not the benefit of, and future success of, their constituents, and of The American People.

Even as this is being written, 87% of Americans do NOT think Congress is representing "The Will of the People". "A Government Of the People, By the People, and For the People" is our charter. Now, without out even a feeble attempt of disguise, it is unabashedly "Of the Party, By the Party, and For the Party". It's a simple matter of legislators who think they are better than everyone else, and that everyone else, the "300+ million" people who are the American Citizens, are too stupid or misinformed to decide for themselves. This present and continuing assault against our Constitution and the visions of our Founders is attempting to extend its control through every one of our Societal necessities.... Food, Energy, Medicine, Finance, Education, Corporations, and even to the extent of monitoring each of our purchases and all that we own. Certainly this type of micro-inspection of each of our lives, justified by "protecting" us, flies in the face of the Fourth Amendment of our Constitution.

"Any society that would give up a little liberty to gain a little security will deserve neither and lose both." - Ben Franklin

LIABILITY REFORM

In countries such as The United States when we find problems, such as the use of certain types of asbestos, which were at one time considered safe, we fix the problem. When a problem is identified, and reasonable measures are taken to change the practice, formulation or design, there is no case for product liability. When, for instance, a medicine is proven to have greater downsides than benefits, or when a medical practice is replaced by a newer, better procedure, there is no rationale for product liability or malpractice suits. If someone is doctoring who shouldn't be, then criminal courts can take appropriate action. Medicines, procedures and courses of treatment all have their own percentage success rates. We must put this back into the front of our minds - that any individual or group of people is neither perfect, nor can ever be, and that this is a natural thing.

When we in the future have created the medicines which easily cure and prevent cancer, HIV/AIDS, influenza, etc., will massive legal actions be taken against those who supply the radiation, chemotherapy, medicines, and surgical procedures of today? What massive legal actions of tomorrow are being organized and planned today? Which of these actions are some individuals already aware? Why don't those same individuals announce that there is a problem now, rather than waiting, plotting and scheming?

"National Legal Care" is a far more palatable (and sensible) way to improve society for all, rather than providing adequate legal support exclusively for the rich, as now. We all recall a famous ex-athlete whose "dream team" rescued him from a likely conviction.

## COMPLEXITY & CLARITY OF LAWS

A law is a concept in the minds of people. It is only recorded in words on paper as a reminder of what the law is. So, if anyone votes for or against a law they do not understand, then their vote does not have any merit. This screams out the ignorance in Nancy Pelosi's proclamation that "We must pass this law in order to see what is in it".

Elected officials must promise to their electors that they will not vote on any legislation that they themselves do not understand well enough to be able to explain in an open forum. Is it unintentional that laws, rulings, and decisions, etc., made by lawyers are frequently vague and ambiguous? I think not. This vagueness and ambiguity ensures that countless billions will be paid to other lawyers who subsequently interpret, argue and appeal (ad nauseum) the merits and meanings of the preceding proceedings, all with hourly rates in the hundreds of dollars. This inducement has yielded an army of underutilized law school graduates. Liability awards, frequently in the millions, have provided added incentives for individuals to pursue this line of "work". Anyone who has his doubts need only watch a few hours of TV.

## STRAYING FROM THE STRAIGHT & NARROW

Politicians and Governments take liberties, even to the extreme of making expensive decisions for "The People", against the people's will, for some political gain or payback of a favor. These events are known as "Corruption".

We must be cognizant of this control by Government, and must exert our rational minds to maintain control of those things on which we spend our monies. It is The People's Right to make these decisions, not a government's right in a representative, free society.

Political arguments often use emotional pleas and dreams of utopia, or guilt from a victimization mentality, rather than rational, logical points of argument, as has been previously stated. It is up to each of us to use our own minds to winnow out the real information from these desperate, urgent pleas. The "voices of reason" are often slow, as they require a time to gather information, and objective consideration prior to making a decision. We must always be wary of those who would try to create an emergency where one doesn't exist, and skeptical of those who chant the party line, regardless of the nonsensical "logic". Our Founding Fathers recognized that people will have various opinions and goals, and that negative forces of religious or dogmatic philosophies would try to gain control of American Government to satisfy their agenda.

The US' visionary founding documents directly and clearly addressed the continuous threat of religious or governmental domination over society, and they insisted on keeping religion out of government and out of excess influence on government. For this same reason, Government should NOT support any social or religious or fraternal organizations. If the Government supports any one of them, whether their mission is religious, social, the arts, sports, "the media", or for any other "social" purpose, it is a gross disservice to the people of this great country. These organizations MUST be supported directly by charity of people who feel their mission is worthy. Those not supported are by definition not what the American People want.

Subsidies and illogical control invariably distort the free market. Our government has spent tremendous resources on the "science" of creating the perception of $CO_2$ as being a dirty Climate Change pollutant, and virtually nothing on trying to disprove that theory. $CO_2$ is one of the only 4 necessities of life. And now, we're on the cusp of smashing industrial productivity, skyrocketing the cost of Electrical Power, on which the price of almost everything depends. This increase in cost of goods and energy will by itself, promote perhaps 50% inflation in a period of a few scant years. Is this merely a way to attack people's life savings? And oh yes, of course welfare payments will need to go up dramatically. Is this "economic equality"? Does this end justify these means? Does this unequivocal discrimination of an entire class of people, the Productive People, yield a real benefit to society? Or does it just suck most people, including the Poor, down to an average lower level? This will of course pave the way to a society with "The Ruling Class", who can vote their own special incomes and benefits, and leave the masses of everyone else poor, but "equal".

UNEMPLOYMENT

We need to spend a moment on Unemployment and "Creating Jobs". The "Unemployment Number", now at about 9%, is a good measure about the health of "Industry", of making things, and of the necessary work which supports Industry. The Unemployment Numbers are a thermometer or gauge of the state of the Economy. It is the success of the goal of having people working which is measured by the Unemployment

Numbers. If Unemployment is high, then it is simply a measurement that there is not a strong need for people to work. Of course this elicits a strong emotional state in a Society, and it is generally reported through a few examples about the desperate condition of those not working and their financial troubles. Ultimately though, the Unemployment Numbers are simply a measurement of the demand for goods.

Just as when we are measuring the temperature in a room with a thermometer, and the thermometer shows the room is too cold, we want to change the things which encourage the room to be warmer. We want to add fuel to a heater or close a window to prevent cold air from coming in. As noted before, when we want to warm a room, we DON'T just put a tiny heater under the thermometer. Of course the thermometer would show a higher number, but THE ROOM WOULD NOT BE ANY WARMER. In this analogy, that is the same as "Creating Jobs" which are non-Productive jobs. We need to un-restrain the natural factors that warm the room. We need to open the shades and let the sunlight in. We need to remove the restriction of airflow into the stove; we need to make the addition of more fuel in the stove. I believe that the popular term "unemployment number" is not an objective, un-biased measure of economic health. A much more accurate number is the "Percent of Americans working". This statistic measures Americans working a full 40-hour workweek, and not those working a minimal amount in order to meet a "work requirement" necessary for a governmental handout.

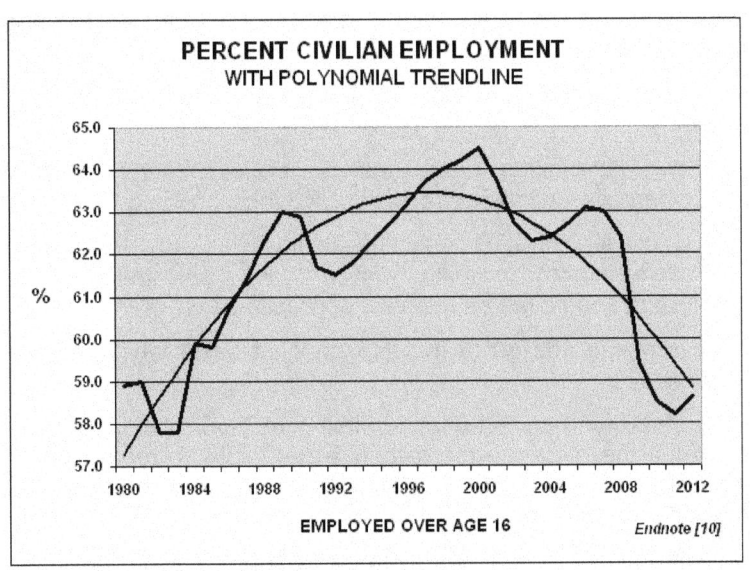

**PERCENT CIVILIAN EMPLOYMENT**
WITH POLYNOMIAL TRENDLINE

EMPLOYED OVER AGE 16

*Endnote [10]*

FOOD FOR THOUGHT

Here's what many say we should be thinking about in fixing America:

Abolishing all Federal Departments and agencies that exist outside the scope of the spirit and intent of the Constitution, and reduce taxes corresponding to these reductions.

Reorganizing all Federal Departments and agencies with the goal of reducing and/or eliminating overlap and redundancy.

Overhauling Federal Pay & Benefits.

Removing artificial restrictions to oil drilling, gas drilling, and coal mining.

Abolishing pending Cap & Trade/Carbon Credit programs and laws.

Insisting that real accounting be used in Federal and Congressional reports.

Demanding fair, accurate and unrestricted reporting from our media.

Passing a balanced budget amendment.

Eliminate "Baseline Budgeting".

Ensuring that "Homeland Security" abides by its original

mission statement, and does not morph further into a watchdog of the citizenry.

Passing new legislation on the legal scope of Executive Orders.

Severely limiting databases on citizens (guns, medical history, emails, cell phone records, GPS locations, etc.)

Simplifying the IRS code and eliminate all provisions which give preference to (or unjust burden upon) a selected few. The tax system should not be used for political means.

Reversing 162(m) of the IRS code of 1993, which capped salaries.

Establishing an "Alternative Maximum Tax". Should any one individual have to pay more than $1,000,000.00, regardless of their income? How much in Federal services can one individual possibly receive?

Returning to a Precious metal(s) Monetary Standard.

Abolishing the Federal Reserve.

Instituting import tariffs and fines appropriate to the Economic/Political situation.

Simplifying congressional legislation, abolishing earmarks, and limiting legislation to the issue at hand, without superfluous issues.

Insisting that all branches of our Government follow and enforce the Law. This obviously includes laws "pertaining to illegal aliens".

Having a national dialogue on the elimination of the Electoral College.

Instituting tort/liability reform, with reasonable caps.

Establishing a watchdog webpage, with daily updates about what our representatives in Congress are doing.

Eliminating all PAC's, SuperPACS, Special Interest Groups, etc.

Instituting Federal funding for Federal elections.

Making campaigning on Federal salaries and funds illegal.

Demanding that our elected officials behave legally and morally. They should be held accountable to a higher standard, and set a moral example as well. Enforcing harsher prison terms for violating officials.

Returning to "citizen politicians" envisioned by our Founders.

Legislating term limits for all members of Congress. Also legislate rules whereby Members of Congress receive pay and benefits in line with the populace.

Abolishing gerrymandering.

Voiding the vote of legislators unfamiliar with the specifics of legislation.

Abolishing the 17th Amendment of the Constitution, which in practice has shifted power from the states to political parties and the Federal Government. *[11]*

"The powers delegated by the proposed constitution to the federal government are few and defined." - James Madison

Our citizenry must behave ethically. Anything else will erode our country into either a police state or lead to social collapse and anarchy.

COLLECTIVES

How can collective medical care, collective charity, or collective anything possibly be less expensive or better? It can't. For some odd reason, many people believe that collective, "public" resources such as transportation, housing, roads, utilities, and on and on are cheaper in a city because so many people are sharing those resources. Or, that it should be cheaper to live in a state with a high population as compared to living in a low-population state. Still, the overwhelming evidence is that in all of the high population collective areas, the cost of housing, transportation, parking, food, entertainment, education, energy, and everything, absolutely EVERYTHING is much, much more expensive, and there is no rationale that it would ever be cheaper than living in a low population density area.

"When we get piled upon one another in large cites, as in Europe, we shall become as corrupt as Europe." - Thomas Jefferson.

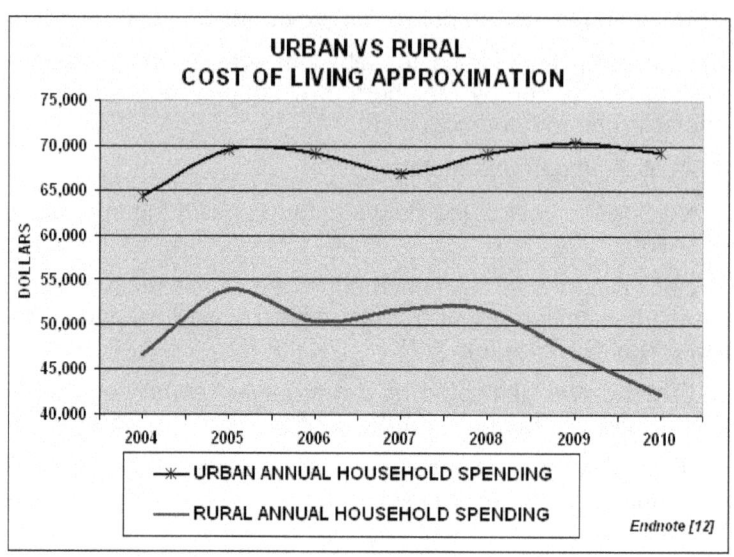

URBAN VS RURAL
COST OF LIVING APPROXIMATION

*Endnote [12]*

In 2004 the average household annual expense in urban areas, was approximately 30% higher than rural areas. This does NOT include any governmental spending on infrastructure, etc. Through 2008, the 30% figure remained fairly constant. In 2009 the variance climbed to 50%, and in 2010 the average household in urban areas spent about 60% more than in rural areas.

In the US, we can look at any major city or highly populated state and see the same result. When we look at the ultimate example of collective everyone, where no one really can own anything, where everyone is paid the same; the average standard of living is much lower.

Look at China, for instance, before the Chinese embraced Capitalism. Look at East Germany vs. West Germany. Look at North Korea vs. South Korea. In every case, "Collective" or Communistic ideologies always have far greater (but equally fair for everyone) poverty. People in collective countries have been "liberated" from the incentives to perform, so collective Productivity is necessarily much lower. How many people other than the "elite ruling party" are living the good life? How many Iraqis starved to death while Sadam Hussein built yet more and more palaces? How many have reasonable medical care, how many have "Kodak® Moments" with family and friends, how many have a bright vision of the future, how many will reach the Super-Productive level of self-realization? This is exactly the farce and failing of Communism. China, although now "CINO" (communist in name only), has embraced and is now totally won over by Capitalism and Productivity. They now value the ability of an individual to excel and be admired for their contribution and the expanded lifestyle their performance has earned. Many millions of Chinese starved to death under Mao Zedong, in the perfect state of collectivism. Now with a growing Capitalistic spirit, how many advances have been made in the standard of living for almost everyone, with access to knowledge, and with monumental achievements of technology? Look at any of the other Communistic-type states such as North Korea, Cuba, and so on. Those people must scrape out a living, or die starving, (except for the leader class, of course.)

We only need a collective when the capacity of Productive individuals is unable to meet a need, such as for Military, highway systems, broad sweeping scientific research, local public education, public mail, and in some cases, public utilities. All other public institutionalized things are ALWAYS more expensive for each individual.

If, for instance, the cost of living below sea level is more expensive than an individual or a locality can afford, then they shouldn't live there. If it is too expensive to make living on top of a volcano safe, then no one should live there, and government must not pay for it, and we should not have our money taken in taxes to pay for it. If living in an area proven to lack long term sustainability for growing food, people should move to an area suitable. If individuals want to grow crops in the desert, the growers must pay for the water themselves. If they can't, they must grow crops somewhere else. It is as senseless as having our tax money used for a heated greenhouse for someone's dandelions to be grown on top of a bare mountain peak in Maine.

We could rationalize that we should take all babies from their mothers when they are 6 weeks old...After all, mothers are too stupid to take care of their children, aren't they? We could show the 0.0001% of children who are mis-treated by their parents, and justify taking all children, couldn't we? Then, we could set up nurseries, kid-care, youth camps and senior camps. Think of the tremendous number of new jobs this would create! All of the care givers, teachers, counselors, coaches, psychiatrists, and so on, which would be needed. We could teach each person exactly what the government prescribes. Then, everyone would start off on an even footing, everyone knowing the same information, having the same morals, and we could instill the beliefs that are exactly the ones the government wants! We could encourage certain individuals to have more babies, and some less babies. We could come up with the ideal citizen. All of our people could be genetically ideal citizens! We could eliminate any thoughts of crimes, any thoughts of un-healthy dietary practices - it would be an ideal society!

Of course Adolph Hitler tried this in Germany in the 1930's, but it didn't work out very well. And no, it's not because he didn't give it a good try. Even if he had today's technologies and computers, it still would not have worked.

Thank GOD Hillary didn't get elected, and didn't succeed in giving a $5,000 prize to each girl for having a baby, as proposed in her 2008 primary campaign rhetoric. Sensible individuals would not consider this "baby bonus" in their decision to have a baby. Perhaps some poor girls would feel this grant would give them some sort of sense of self-worth. Certainly a girl who does not have the sense to realize that $5,000 will not even feed a baby, much less clothe it, provide for its medical care, and its rearing costs could not have enough sense to teach a child the morals and ethics required to be a Productive person. It's just a reward to have a baby. The biggest outrages against societies, those responsible for genocidal crimes, all start out with controlling media, conviction/imprisonment of the "thinkers", the philosophers and artists, scientists and teachers, and honest public figures. We can see all of these things in human history, from the Romans to the Nazis, to the Socialists and Communists. Those abominations of societies all started out with the attempt to control thought and expression of thought. Although most of these campaigns also generally excluded free expression of religion, the most heinous times were controlled by religions, such as "The Spanish Inquisition" by the Catholic Church. Fortunately, Catholic leaders saw the error in their ways, and opened their eyes to the rights of individuals to think and act as individuals.

RELIGION

In stark contrast to some other countries, we have consistently valued "The Separation of Church and State". This likewise applies to separation, from our government, of groups who have religious fervor for social issues. Any of these groups are of course allowed to peaceably bring their concerns to the public's attention, but it should not be allowed for those groups to become imbedded in government. It's sensible to deny any "Special Interest Groups" from having disproportionate representation or control in Government.

Religions are mainly responsible for helping to develop and maintain the morals of a society, be they "good" or "bad". Since religions are all different, some believers see other religions as having immoral aspects. Christians, for instance, do not agree with tribal practices of cannibalism and sacrifices. Christians generally believe in the equality, and special roles of both men and women in society. Changing at present is the Christian belief that a "marriage" is only valid with one person of each gender. Ample evidence exists to show some religions hold the belief that if they are not condemning the beliefs of another religion, that this tolerance is in fact the same as condoning those beliefs. If, for instance, Christians are not sending missionaries into other countries to try to change their beliefs, then the Christians are guilty of supporting those beliefs.

Many religions have sent their missionaries in an attempt to change the ways of other societies. Those societies, having sovereignty as "members of the world" have the right to choose their own morals. If they have studied and understood the rights of individuals and groups in other societies of the world, so much the better. It's optional for "outsiders" to inform them of rules other Societies practice, and perhaps explain the rationale, but not to use psychological, financial or manipulative techniques to change them.

For instance, HIV/AIDS is spread through sexual contact, "dirty needles", and other unsafe activities. It is each society's moral obligation to make this information available to their people, so that the people can make informed decisions on modification of their behaviors to contain this epidemic. In at least one society, men believe that sex with a virgin will cure HIV/AIDS. It is up to that society to change their level of public discussion to explain this belief as totally wrong. Outsiders can help inform the leadership, but it is up to that leadership to make the changes. Of course if the local government tells outsiders to stay out, then so be it. Let the outsiders drop their "holier than thou" attitudes, and respect their right of sovereignty.

This all stems from the self-approval process that makes people subscribing to a belief feel superior to people who have other beliefs. Demonstration of beliefs has led to social practices which others consider truly evil. In the US, we can easily see what we would consider to be not religions, but rather Cults. Note the practices of Jimmy Jones, and David Koresch, and others reported in the contemporary media. Christians as a whole do not now believe in the overwhelming control of all individuals, but rather allow gentle interactions among members to set standards, rather than the unquestionable edict of senior members. A church in the 1400's (which literally controlled governments) amassed too much power and went to the extent of torturing and killing people merely accused of thinking unfaithfully. Religious leaders in Salem, Massachusetts got themselves into a similar evil predicament in their literal "Witch Hunt". In this case, those leaders believed that GOD was punishing their community due to their lack of condemnation of people with magical powers. - A case of one part of society blaming another for natural climate changes, which affected farming.

We are all very proud of "The Pope of the Catholic Church" and "The Vatican" for the recent global apology for immoral deeds by a few church officials. We are proud of the Catholic Church for rightfully accepting full responsibility for the negative actions of its members in this way. This has not weakened the image of "The Catholic Church", but rather strengthened it considerably by developing increased respect from those both inside and outside of the Church.

The spoken and sometimes unspoken term "social good" is referred to frequently in religious and other works and in enduring social commentary, as are the moral prioritizations of: 1) family/religion, 2) community, and 3) country. In some societies, these priorities have been in another order. In Germany in the 1930's, the priorities were in the order of: leader, country, community, family, then finally religion. Children were instructed to report their parents for thoughts or words against Hitler. Most, but not all people currently believe that what happened in Germany in the 1930's was evil. It is a sad commentary of our time that individuals, communities, or even leaders of countries should promote this blasphemy and the political theories that promote it, and to deny "The Holocaust". Fortunately, General Eisenhower and others recognized the importance of documenting for all time this horrific attempt at genocide. Holocaust deniers of today surely must have a vile agenda themselves.

PEEK-A-BOO, BIG BROTHER SEES YOU

A new technology, perhaps of some benefit by allowing energy users to measure how they are personally using energy and therefore re-prioritize their needs, is the "smart meter" for electricity, which is an online based unit. How soon will this technology, which, it is implied, will allow each of us to "manage" our own usage, actually be used as a tool of government to control us? Will the government watch the operation of all of our appliances and demand how they are to be used? Or will the government directly control, without our input, operation of our heating and cooling, our entertainment, and other aspects of our lives? How many computer viruses will infect these systems? How many people will die from lack of cooling in the summer or from lack of heat in the winter? And, the biggest question about this type of direct monitoring is the age-old question of how much tracking of every individual's personal lives is OK? Even in a "New World", with terrorism against freedom, is the monitoring of every person justified to avoid a discriminatory practice of tracking only the possible terrorists themselves? Wasn't it Nancy Pelosi who said we must regard all returning Veterans as potential terrorists? What would ever make her believe this?

If people individually want their health records available on-line, they can "opt-in", and pay the actual cost themselves, or better yet, carry around a memory stick with all their medical records on it. Of course there will be one or two individuals who will benefit from "on-line medical records", but at a great loss to the sanctity of the "Doctor-Patient Relationship". Could the posting of all Legal Records on-line, which would violate the "Lawyer-Client Relationship", be next?

And, who are we now considering to track as potential terrorists? In a recent document from the Department of Homeland Security, this list includes returning soldiers, conservatives, and many others. Should it also include our elected officials, teachers, union members, farmers, and all Productive people? Should modern patriots and people exercising their Constitutional Rights also be considered as potential terrorists?

CAPITALISM - A RARE COMMODITY

We only need a cursory overview to see how Socialism, Marxism, Communism, and other social disorders have stalled and suppressed real progress, that is, improvement of the average standard of living and enjoyment of people of those societies. A perfect example of these failings is the nationalization of Industry and it's supporting institutions. We can look at other societies who have tried these strategies (Denmark, Canada, UK, and Greece, for instance), and we can easily extrapolate the experiences we will have as our society continues its leftward decline.

Dreaming about everyone having perfect everything, the same of everything, or even parts of this dream, is a sirens' song. Involvement in this emotional foray drains away vast mental resources, which rather should be directed toward creating greater Productivity. ALL great things are resultant of Productivity. All of every physical thing we have, and hope to have, results from Productivity. All food, housing, transportation, clothes, medicine, tools, and other things with actual value come exclusively from society's net Productivity. When the people of a country organize themselves into cooperative Productive groups, that is "Industries", with Incentives proportional to individuals' Productivity, imaginations and attitudes soar, leading to greater and greater Productivity. Greedy governments or other organized "regulatory" institutions readily stifle Productivity through taxes, fees, fines, regulations, and so on. There are but a few essential regulations required to maintain "fairness" and natural competition in a Free Market.

We're seeing more and more scheming away from people's responsibility for their own upkeep, their housing, and their food. "Universal Healthcare" is but one of these socialist maneuverings. Do some believe we should guarantee happiness? These are idealistic, and not practical, or achievable goals. They're dreams of utopia. These things can only ever work "on paper". This state of perfection totally ignores human sociology. People need to have incentive and challenges to lead them to Productivity. Without Productivity, there will be nothing for anyone, regardless of how much money is printed. Sensible people know that if we give children everything they want, they will be spoiled rotten. Why do we think that taking away the reasons for working should affect our society as a whole any differently?

If we continue our present course, we will in hindsight, be able to clearly see what we are now going through. Germans became very disturbed socially under Hitler, and couldn't see what was happening and were unaware at the time how their thoughts had become anti-social until after that dark moment of history had passed. During the "Witch Hunts" in Salem, Massachusetts and during "The Spanish Inquisition" people had generally convinced themselves they were "In the Right", but later realized how absolutely brutal and wrong they had become.

We can see what happened to Jews and others under Hitler in the 1930's and 1940's. Look at enslavement of Christians in the Ottoman Empire. How about the "Churches" of Jim Jones or David Koresh? All the above-mentioned groups were manipulated and their senses of reason blinded by strong emotion.

In a Communist society, with all ownership regulated by the state, everyone owns everything, and at the same time, nothing. Even in the most "pure" of implementations, there are always the "ones in charge", who say they deserve more and get more compared to everyone else, since they believe they are superior to "normal" people. In Socialist countries, everyone gives to everyone else, but a person is allowed ownership to a degree. The most productive societies, and therefore the ones with each individual producing more, are Capitalistic without exception, due to the "carrot factor" - incentive. Some interesting societies did at one time in the distant past enshrine the values of Productivity and Science, but then somehow lost direction and wound up with a large population of poor and very poor. Some of these societies have made tremendous income selling minerals to capitalistic nations. In these same minerals-only countries, there seems to be a strong desire amongst the "rulers" to keep most people uneducated and uninformed - to use them as virtual slave labor. It is truly a crime when the global price of those minerals gets high. These party/peasant societies then have more than enough money to build huge palaces and huge armies. For example: Iraq under Sadam Hussein. Literally thousands died of starvation whilst Sadam continued to build yet more palaces.

This begs the comparison between Capitalistic and Socialistic societies, and for that matter, Communistic societies. If everyone deserves and gets the same basic food and care, do the poorest have a better life than the poorest in a Capitalistic society? No. In a Capitalistic society the true of heart give charity to those who are truly in need. Those in need will certainly be better off due to the fantastic riches of a Productive Capitalistic land, and have far greater opportunities for employment. This is after all, the crux of the matter. What is more important to equality? Does everyone starving equally make a better society than one where the poor have plenty to eat, but others in the society have several plasma TV's and nice cars? This is a normal distribution and represents the spectrum of Productivity in a free market society.

The best society must surely be the one with equal opportunity to a decentralized education, with education of both boys and girls as key to their future. We should support the values of local public supported education to allow the poor an equal opportunity in "pursuit of happiness". We can not guarantee that happiness, and must not remove incentives to be Productive members of society. To produce more goods enables an acceptable standard of living for the poorest of us, with enough food to eat, and a place to make a home. Even a log shack with a garden spot to grow a portion of a family's food is enough for a loving home.

We must of course recognize the fact that some contribute far more to society than others. We must allow the enjoyment of our earnings to be in proportion to our contributions.

If one member of a tribe is catching more game, shouldn't that person be recognized as being special, and get more of the best meat? Shouldn't their prowess be recognized and rewarded? Maybe that especially talented person should be allowed to choose the most desirable mate, to get fat, and to enjoy more of the luxuries that society offers. Won't the passing along of these superlative genes cause ever-higher productivity for future generations? Of course it will. Natural Laws regulate "the survival of the fittest" and there is no departure from that in this instance. The genetic characteristics promoting superlative hunting skills are passed on to children, as is the parental nurturing and societal respect for good hunting skills.

The whimsical dreams and ideals of a "One World Order" or "singular world religion" or a "one world government" have popped up variously around the world, throughout recorded history. ALL attempts at these things have cost countless lives and fortunes of societies. Now, all of a sudden we have computers, and the concept of a "one world order" will magically work? It has never, and can never work, as the inevitable repression of groups leads to revolution. Look at "The Roman Empire", Genghis Kahn, The Ottoman Empire, "Imperial" countries like Britain, Japan, and most recently the Soviet Union. Both achievers and peasants alike are forced out of their freedoms and their earnings when such a thing happens. A One-World Order emotionally seems a good idea, but every person is an individual, and each person has individual needs. Should everyone have to eat 8 oz. of tofu per day? Should everyone be within the "healthy weight" category? Of course not. Some have made movies showing that if everyone is an exact clone of all the others, it can work. We are not clones, we are all individuals, and even identical siblings are individuals.

## PATRIOTS

Today we have true Patriots in almost every aspect of our society. Sensing an increase in the angst and gnashing of our society's collective consciousness, New American Patriots are continually finding themselves. These American Patriots rarely act for their own gain, but rather for the betterment of the condition of our country. These Patriots are not groups trying to stir up controversy to maintain the validity and purpose of their mission, to embroil themselves in the ever popular "mission creep". They are people of genuine concern and who are willing to use their own resources in their quest. These Patriots are not paid members of a centrally organized socialist group, as were the "Occupy Wall Street" group.

As citizens we MUST pay attention. Since it is becoming increasingly obvious that the Media has an agenda of their own in the political arena, it is up to us to seek out additional sources of information. We must educate our children regarding what WE believe. We cannot afford to allow the Media or the Federal Government (via Federal Education programs) to dictate to them. We must research and be vocal with our findings and beliefs.

There is nothing new here. There are, however, new opportunities to regain control of our Government and our lives. The Internet has opened a new door for the citizenry. We can seek the truth, and express our opinions to a larger audience than once thought possible. It is up to us all to DO IT. If not, there should be no doubt that we will be displeased at the consequences our own inactivity.

## CONCLUSION

Our founders knew about governments too strong for their people, and about the rebellions and civil wars that brought countries back to sustainability. They knew of the evil emperors, kings and queens. They knew of countries run by religious zealots. Our founders knew as well of ruthless dictators and imperialists. They knew the necessity and history of the Holy Crusades. We have allowed our government and society to evolve into an unsustainable dream. Small transgressions, often unnoticed, keep occurring. And now we are in a state that is far removed from its original intent.

Our American Heritage and our American History are important and essential topics that should be taught to every school child. That History is of course the real facts, not book burnings, not emotional manipulation, and not 'rewriting history' to push social re-engineering. Negative emotions distort each of us in of our ability to make informed decisions. Our decisions made not in haste, not through feelings, but in a practical direction employing "the voices of reason" are the good decisions that pass "the test of time". Herein is the essence of the cure to the excessively negative "Politics". It is the voice of reason, itself.

# SECTION 2 - MAN AND THE ENVIRONMENT

# FOREWORD

There is NO proof that increasing $CO_2$ (or any other Human influence) significantly affects Global Temperature.

It's "Global Climate", "Global Cooling", "Global Warming", or "Global Sea Surface Warming", or "Ocean Acidification", or "Global something". Of course the climate is changing. Of course Global pH changes over time. The only planets that don't experience change are DEAD PLANETS. Some believe there is no relationship between Global Temperature and $CO_2$. This belief is wrong. If we take the raw data from the infamous graphs presented in the movie "An Inconvenient Truth", and examine the Global Temperature and level of Atmospheric Carbon Dioxide, it is absolutely clear that there is a relationship between the two. The crux of the matter is what this relationship represents. The $CO_2$ "crisis" is a collective emotional disorder. We are being led to fear that humans, particularly through the use of ancient stored sunlight, are pulling the trigger on a massive calamity of unimaginable scope. It's being said that we're unwittingly facilitating our own demise and the loss of all we hold dear.

This "crisis" is also being manufactured through manifestations of hate. There are those who hate the success of others, the successes and greater wealth of those who have followed their dreams and organized the efforts of others into productive enterprises. And yes, the $CO_2$ "crisis" involves misguided charity. There are people and groups who want to milk the $CO_2$ "crisis" for all the money they can, to win favors for those "less fortunate". Truth be known, many of these "less fortunate" are in these situations not due to lack of resources, but due to the repressive governments under which they struggle to survive. If massive funds are taken from governments as fines for use of fossil fuels ($ Quadrillions) will those "less fortunates" suddenly be more prosperous, or will the money simply line the pockets of those repressive governments? What has Iraq done with all of their massive amounts of money? How many palaces have they built? That country must improve from within to allow property rights and family savings outside the reach and erosive powers of governmental influences.

Considering the Science of Climate Change we must accept that the climate changes without human influence, at times in rapid and massive ways. We have found that average atmospheric levels of $CO_2$ and average global temperatures are somehow related. We do NOT have a clear indication from ancient proxy data that a change in $CO_2$ causes a change in global temperature. We have several very well produced "Science Documentaries" which claim the "snowball Earth" was melted by volcanic discharges of $CO_2$, and hydrogen sulfide, sulfuric acid, and other gases. Some also claim that methane released from undersea hydrates spontaneously ignites when it hits the atmosphere. If this is true, why do we need pilot lights on stoves and furnaces? So although well made by filmmakers and their "consultants" their science falls short of correctness and impartiality, hallmarks of the "Scientific Method".

What, then, is the Relationship between $CO_2$ and Temperature? There are seemingly only 2, or perhaps 3, rational theories:

1. Increasing temperature causes an increase in $CO_2$.

2. Something causes both temperature and $CO_2$ to increase.

3. Increasing $CO_2$ causes an increase in temperature.

We can rule out 3 as a false theory since there is currently no correlation with contemporary measurements.

(Further information can be found in the book:

Global Warming for Dim Wits: A Scientist's Perspective of Climate Change [James R. Barrante])

As we shall see, measurements on Mars prove that $CO_2$ has negligible effect on temperature.

Neither of the two remaining theories implicates humans or underground coal mine fires or $CO_2$ from volcanoes as having any effect on global temperature or climate.

Humans are otherwise planning and doing some things, however, that could potentially affect climate:

Removing plant life from large areas of the Earth, exposing soil, and reducing natural transpiration of moisture into the air.

Building large cities, which has the same effect as 1., above.

Large scale installation of things that slow the wind (wind turbines).

Large scale exploitation of ocean currents and oceanic temperature difference for energy production.

Large reflectors in space which focus more energy on the Earth.

Draining swamps and inland seas, increasing land surface

area.

One mission of this book is to convince the reader that these important topics deserve rational consideration by the world's citizenry. We must remember that politicians and governments and the organizations who support them (media, UN, unions, lobbyists, etc.) are not necessarily "pure of heart" and are frequently motivated by their own interests, not in the best interests of their constituents, members, or even of humanity itself.

By believing, at face value, claims of human-caused environmental damage without the necessary un-biased evidence, we are shirking our responsibility for the oversight of the future of our lives, our standard of living, and our planet.

## INTRODUCTION

The obvious importance of the maintenance and improvement of the quality of the Environment is that a healthy Environment is necessary for the basic survival of humanity. For any number of reasons, many individuals and groups try to sway public opinion to their belief(s) that they have the solutions to all Environmental problems, and, therefore, we should follow their lead lest we perish.

# WHAT'S THE PROBLEM?

There are 2 types of "Climate Change Deniers". There are those who deny the Climate is changing, and those who deny Climate Change is due to forces outside the reach of human influence.

When did the Climate start to Change? Was it 50 years ago? 200 years ago? Don't our Scientists say the Climate has been changing for literally Billions of years?

To deny Climate Change is to deny the vast swings experienced by Earth over the millennia, and to deny the very history of our planet. Indeed the Climate is Changing. Changing as many Billions are being spent on trying to find a relationship between humans and Climate Change, even though there is no defensible link. Very little money is being spent to prove the opposing view, the view that humans are having no measurable effect on Climate. There are no changes in amplitude or rate of change of Climate that are coincidental with "The Industrial Revolution" as compared to the pattern of changes in the Climate Record. "Climate Change" is generally presumed to be an effect of weather change from temperature change. It is now becoming very clear that Human liberated $CO_2$ has not changed Global Average Temperature, either by "Global Warming", or "Global Cooling". Perhaps Human $CO_2$ has had an effect of maintaining a "Global Constant Temperature", in effect paralyzing the otherwise natural wide swings of Temperature evident in The Geologic Record. In fact, the most recent 1000 years has been uncharacteristically stable. The most recent 1000 years includes much greater Human Progress, Productivity, and advancement than during any time prior.

Asking the question of whether a stable Global Temperature is good for "nature" and "the environment" or not is meaningless, since this topic is well beyond the scope of responsible speculation. Certainly, without the sporadic and extreme events in the Earth's ancient and recent history, we would not have the same flora and fauna of the Earth we have today. We know 99.9% of all species which ever existed are now extinct. With less frequent and less violent Climate Change, maybe some of those things now extinct would be alive today. Maybe Humans would not.

MOTIVES

There are individuals with misdirected passions about our Environment. Individuals who would try to convince us that the one aspect of our Global Society - the aspect of Production, is rightfully, should be, and will be further attacked, taxed, and drained. Those misdirected persons, no matter how sincere they may be, have mistakenly linked natural changes on our Earth, natural changes which have occurred many times over millions of years, to some sort of guilt over the blessings showered on our Global Society.

There are others who are far from being misdirected. They know full well what they are promoting. Their excuse for their irrational attempt to pillage the ever shrinking sectors of our Society which are still Productive, is the Unproven Theory that human emitted Carbon Dioxide is changing the Climate, and we'd better slam a conviction on energy users, "just in case". We're not talking about a fine of millions or billions of dollars, here; we're talking many trillions. This is where the Politicians and Lawyers get passionate and motivated. What better source of "new, under-tapped revenue" than the very section of Society which creates ALL real value? They won't go after those who only absorb, even those who already absorb vast resources. That's akin to drilling a dry well. They attack on behalf of Society, they convict, and they condemn. They suck the very wind out of the sails of those schooners that made our country great.

There ARE REAL environmental issues today; Issues such as garbage killing our precious animal friends, filth and lack of sanitation, unsustainable stripping of valuable eco-forests, and other crimes not committed by our Industries or our Free Market System. Other REAL issues include extinction of species by poaching or habitat destruction and destruction of ecosystems by both intentional and unintentional transportation of "invasive" species.

The biggest problem today is the lack of focus on the REAL issues. Why this lack? There are those who are not true of heart for the Environment, who are using Environmentalism as a pawn in their scheming. They're scheming to leverage away vast monies from the Productive Sector. How will you like being thrown under the bus amid a droning of propaganda once "Their" plans are realized?

A significant aspect of today's fear of climatery fluctuations is easy to understand. It is very easy to scare a large number of people, particularly with today's instant communications. It is rather difficult, however, to "un-scare" people in order that they think something is being done in the "right" direction, and it is really immaterial whether or not those changes have any real effect at all. It's just "well, at least we are now doing something about it", and it doesn't matter if even what we are doing is in the exact wrong direction, making things even worse. Who would ever really know if we had made things worse or not?

Many people, the "Global Warming Advocates", are correct that the Geologic Record shows that high Global Temperatures have existed at about the same time as high levels of Carbon Dioxide in the Atmosphere. They contend that the high level of Carbon Dioxide is responsible for the Warmer Climate. These people, Al Gore and his ilk, cling to the theory that an increase in Carbon Dioxide causes higher Global Temperatures, and that Humans, and particularly Human use of Fossil Fuels, is responsible. Case Closed.

This strategy of persecuting Fossil Fuels as Environmentally damaging began with the United Kingdom's Prime Minister Margaret Thatcher, who desperately wanted to escape the dependence of Britain on Fossil Fuel energy supplies, since:

The coal mining Unions had become too powerful and had already held the country hostage to their demands, and

The Oil Embargo by the Mid-East in the early 1970's proved Britain could not depend on those energy supplies either.

So, Ms. Thatcher set out to fund scientists who could prove (and not fund those who would dis-prove) Global Warming was resultant of man-made $CO_2$. In this way, Ms. Thatcher wished to justify building many more nuclear power plants, to keep Britain insulated against unreliable energy supplies.

We're currently being inundated with media reports espousing the feeling that rising $CO_2$ levels signal a trigger for a calamitous breakdown of life on earth. Of course, the doomsayers are financially rewarded for their efforts in convincing "the public" about dangers of $CO_2$. Those proposing a neutral or beneficial effect of increased $CO_2$ are cast as villains of humanity and of the environment.

The beneficial effect for the doomsayers will be the collection of massive taxes from industry, and redistribution of that money to likely accelerate population growth in the poorest countries of the world. And it will also increase the number of poor in developed countries through improved medical care and access to adequate levels of food. When will these people be free from their bondage of charity?

Of course any scientist who so much as hints a climatic change has anything to do with something other than humanity's evil actions will be totally stripped of any status in their field. Not at all unlike the "witch hunts" or "The Spanish Inquisition", You get rid of a witch and things get better, or maybe things don't get better, and you need to drown more witches or to sacrifice more virgins. Eventually things improve, and, it is implied that you were therefore successful.

It was a shocking announcement by the US Environmental Protection Agency that $CO_2$ is now listed as a pollutant. I would have loved to have been a veritable "fly on the wall" during their internal discussions on this topic. I'm sure their conversations would make a great documentary of intrigue and manipulation. This move also directly implies that less $CO_2$ is better for the Environment, in the same way as is less lead and other heavy metals in our drinking water. Before someone says otherwise, however, there is no difference in the chemistry of "Man Made $CO_2$", and "Natural $CO_2$". You could not grab a molecule of $CO_2$, for instance, and trace its origin, except possibly through measuring isotope ratios. Plants, animals and other forms of life respond identically to "Man Made" and "Natural" $CO_2$. In fact, all carbon is from nature, and it cycles through the environment in various forms. All life on earth is described as belonging to the category of "Organic Chemistry", which literally means Carbon and something. People are just not making any new carbon.

By the way, why did we recently shift from accusing humans of perpetrating "Global Warming" to "Global Climate Change"? Because in 2003, even though the atmospheric $CO_2$ level continued to increase, Global Warming reversed to cooling. So, if evidence of Global Warming went away, we can still accuse humans of something, even if it turns out to be "Global Cooling". (Or changes in global weather, more drought or more flooding, or more of anything which may or may not be bad). I'm sure "the powers that be" are using this strategy to "keep their options open", and not suffer a defeat "in the public's eye", as the thing they were championing against, Global Warming, naturally reverses.

If we had been in a period of Global Cooling and Climate Change over the past few hundred years, all of this same fervor would still be alive. Although there would be slightly different theories, they would still be directed at those who are Productive, those who generate all real value.

This change in terminology, from "Global Warming", to "Global Climate Change", (and now "Global Sea Surface Temperature Change") although seemingly innocuous, shows an absolute resolve to convict humans, and in particular Industry, of change in the climate, regardless of what that change is, regardless of what un-biased scientists eventually find to be the source. These sources may well include solar output change, Earth's orbit, changes in the amount of dust in the Atmosphere, variations in internal heating, etc. The "forces at work" have made a political decision that "humans are guilty" and they must pay, and the politicians are busy rationalizing that conviction. Lawyers globally are tittering in excitement about this exciting new source of revenue. 30% plus "expenses" levied on trillions of dollars is indeed very big money.

So, the "Environmental Movement" mission statement has now changed to "Stop Global Climate Change". This is surely a much broader mission statement, one that can virtually never be completed, and assures a cornucopia of activities, accusations, and litigation. Even with our almost unlimited data collection and analysis, Global Climate Change due to Human Activities can be neither proved nor disproved. It will be a case of who has the most money, and which outcome will benefit activists the most. Industry and Productivity creates all the actual monetary value in society, and those now in control want to grab more of it and control all of it.

Are we reveling in our arrogance at the thought we can control natural climate change? Isn't this just another step in the attempts at "social justice"? Once fully implemented, environmentally based mandates will never be rescinded for any reason. These mandates will survive even in the face of overwhelming scientific data disproving their worth.

The key to the success of reversing the current hysterical condemnation of Industry is litigation. If those hurt by restraint of trade or artificial escalation of the price of their purchased goods bring cases to court; damages caused by pseudo-scientific terrorism against industry will surely be compensated. Of course Al Gore is not the central character in the present anti-Production and anti-Progress "Environmental" movement, but he has positioned himself as an expert in the field. Alfred Nobel would surely be surprised by such anti-Industry recipients of his legacy.

Even now, we have a federal decree that we must replace many of our incandescent lamps with fluorescent (mercury) lamps.

What will become of our Easy-Bake® ovens? [13] How will our children cook their delicious muffin cakes? Will this supplier provide "infrared heat cooking spheroids", which look a lot like a regular bulb?

UNITED NATIONS

If the real purpose of the United Nations Climate Change Council is just to get money to help "Underdeveloped Countries", to help them control their populations, to help them have improved access to food and medical care, to improve their lot in life, then they should say so. Of course they have said this is one of their missions over many, many years. But, I guess it has not worked as well as they would have liked, in terms of receiving more charity for their causes. But now, if the UN can attach itself to the income of the MOST Productive and profitable Industries in the World, we're talking really, really huge money, in effect purity of Marxism.

Now the UN focus is to convince people that weather is worse that it's ever been, and that Industry is guilty, regardless of the real, long term changes in weather, and whether those changes are natural or "human caused". We of course have more records of bad weather now, of floods and droughts, of terrible hurricanes and typhoons than we have from the past. We have more recorded weather events as we have better data communication and more "history", than before the present "Information Age". Our scientists tell us about the evidence of massive extinctions before recorded history. Although these things are not affecting us right now, a sea level change of a few inches or even a few feet or even 10's of feet (totally natural as evidenced in the geologic record), would seem much worse than the comet which wiped out most life on earth some millions of years ago. When we look at our Moon and other planets in our Solar System, however, we can easily see that massive impacts are normal events. On our precious Earth, most of the evidence of the tremendous number of impacts has weathered away. How easily do we convince ourselves, "Oh, our Earth is special, those things can't happen here"?

CARBON CREDITS

It makes Economic sense to use our resources efficiently. "Fossil Fuels", (literally $CO_2$ being recycled in the "Carbon Cycle" through geologic processes) are a convenient, inexpensive form of energy, of which we have enough supplies for centuries to come. This doesn't mean, however, that we can afford to be wasteful of these important resources. At the same time, by grandiose schemes of taxation and restriction, we are punishing particularly the poorest of the poor, and are putting the essentials of their lives out of reach. We are eliminating many opportunities for the poor to obtain good, productive jobs. We are increasing the costs for heat, for food, and the costs for transportation.

The hot button for this age is energy, and more specifically, fire. In the broad sense, "fire" includes all energy sources dependent on combustion of fuels. This includes internal combustion engines, electricity produced through coal-fired generators, propane fired heating and cooling, and the vast majority of energy sources we are all dependent upon. The "Carbon Credits" program will penalize most people economically, for an arguably false promise about making something better for our descendents. The feeling is that even if this, which causes our further economic malaise, has no effect on the climate whatsoever, then "at least we did something". And, if what we do, (the Carbon Credits system) doesn't do anything, then some will propose that we do even more, prohibiting ALL fire, for any reason whatsoever. So, we'll have nothing but nuclear, wind, water, and solar power, charging batteries for all transportation, and of course aircraft flight will have long been banned in favor of electric trains. All cooking will be done electrically, all heating by "heat pumps". And, with luck for the proponents, the climate will reverse (same as it would have during any of the climate's many cycles with or without people), and then the proponents will surely be heroes, in the same way as they would if they had sacrificed a virgin to a deity and the problem naturally reversed. Due to the lack of recent coverage in the Media, some believe that the Carbon Credit/Cap & Trade programs are a dead issue. This is far from the truth, since the power Industry is already making very expensive changes in their operations.

What if basically outlawing ALL fire of all types reduces our $CO_2$ level so low that farmers have a difficult time growing their crops? Those farmers would have to buy "Carbon Tickets", to have $CO_2$ made and delivered to them, so that their crops would grow. This is a whole other side to the Carbon Credits program. Sell Carbon Credits at ever increasing prices until people don't emit any $CO_2$ at all, then generate and sell $CO_2$ to the farmers - get the money both ways.

Does anyone seriously believe that other countries will buy into any Carbon Credit program? Have other countries enacted pollution laws similar to those already in force in the U.S.? Are we ready to initiate sanctions or hostilities against these countries? The air quality index in third world and in developing countries should give us a clue.

It was a sad day, indeed, when the US EPA made a ruling that $CO_2$ is a pollutant. This was an utterly ridiculous political decision. This was a political decision designed to get money from Industry for massive social reforms. This is an unnatural restraint on Industry, ripping away their rightfully generated resources. This is clearly a case of the socially questionable "Ends justifying the Means".

# THE CASE FOR POLLUTION - IT'S A NATURAL THING

## PUMP IT OR LEAVE IT - IT JUST KEEPS COMING

A "hot button" item in our past was the wreck of an oil tanker, the Exxon Valdez. Although tragic and ultimately avoidable, it was only 1/4 the size of the worst oil tanker spill incident in history. Now, we have "double wall" tankers. Human related crude oil release in the environment is estimated to be about 30% of the total released into the environment. Very recently, vast increases in the estimates of natural geologic seepage may well reduce this "human-attributable" component to be a much smaller percentage of the total. These natural seeps from fossil oil are not the total of the "natural" sources. Rotting vegetation in the bottom of streams, lakes and rivers creates and releases thin films of oils on those bodies of water. In fact, any stagnant water in these areas invariably has a thin oil sheen on it from natural sources.

Many records describe drilling oil wells, and "hitting a gusher". The internal pressure is so high that natural seepage into the Environment is normally experienced before relieving those immense underground pressures. Other records recount various ancient peoples using tar and other naturally concentrated products of crude oil that seeps out of the Earth. One man's pollutant is another's "black gold". How about the "Tar Pits", known to have trapped dinosaurs? That is crude oil, forced by pressure up to the surface, the lighter fractions evaporating into the Atmosphere, leaving behind the "sludge" tar. Oil seepage and "Tar Pits", etc. are natural pollution. In the Gulf of Mexico, there is an "Asphalt Volcano" which has been growing for millennia.

Humans, by their activities, are inadvertently decreasing natural pollution by significant amounts. Humans are decreasing the natural pollution from seepage of crude oil by pumping it out of the ground and reducing the pressure, which causes natural leakage to decrease. We convert this crude oil into many useful products for society, such as energy, plastics, and medicines, which are on the whole, very clean. So, pumping oil and gas out of the earth actually cleans up the environment, although only as an unintended beneficial byproduct of our technological advances.

I'm sure that the oil that Jed Clampett found bubbling up out of the ground stopped bubbling up after oil wells were drilled and the pressure of the deposit was relieved. This is a bit akin to draining a subcutaneous cyst from the Earth. Humans may well be reducing the sum total of this natural seepage pollution by quantities significantly more than all of the oil tanker wrecks in history combined.

But, who is interested in reporting this? Who would ever dare give a positive credit to Industry, when a few really loud voices are doing everything then can to send us back to "The Stone Age"? All while many politicians are doing their damnedest to try to take credit for the miraculous improvements in our average standard of living (which Industry created) for themselves. Those politicians are trying to control Industry, which creates the only real value: "Goods". And the politicians are trying to tax that value. It's NOT "The People" who are trying to tax Industry; it's the politicians. The politicians are trying to tax this value to spend on various unreachable dreams of utopia.

CARBON DIOXIDE, FRIEND OR FOE?

The original massive amount of $CO_2$ in Earth's atmosphere that was converted by plants and animals into "fossil fuel" is being released bit by bit through combustion back into the environment for another cycle through the process. Before those plants and animals captured the $CO_2$, and converted it into the carboniferous remains that turned into oil and coal in the earth, there MUST have been a MUCH higher level of $CO_2$ in the atmosphere, as evidenced by the tremendous known reserves of fossil fuels. Of course, even as this book is written, carbon is being sequestered into more oil and coal by plants and animals as the "fossil fuel of the future". The higher the level of $CO_2$ in the environment, the faster the growth of plants and animals, and therefore the higher the rate of creating new fossil fuels.

Also at the time of this writing, the United Nations IPCC committee still maintains its position that "human generated" $CO_2$ is damaging the climate, which they say will force a terrible calamity that will at some point trigger an apocalyptic disaster for all life on earth. With 52 scientists behind their position, it really seems like an army of support for gloom and doom and a conviction of "fire". There are now more than 700 scientists, all of international renown, who have stated their disagreement with the IPCC position, but the 700+ seem to have a voice like a whisper in a hurricane.

We DO have real, scientific evidence of much higher levels of $CO_2$ in Earth's Atmosphere in the past, which did not result in a calamitous breakdown of all life on Earth. High levels of

Environmental $CO_2$ actually resulted in the fantastic development of life, as we now know it. High levels of $CO_2$ in the Environment were directly responsible for the phenomenal growth periods of the Carboniferous and Jurassic periods.

When we hear the IPCC and other "gloom and doom" predictions of worsening weather, it is certain that apocalyptic changes in climate have already happened many, many times throughout geologic history, even before people "first discovered fire". The IPCC argument goes: "The $CO_2$ is causing warming, the ice caps are melting, and the oceans are rising."

Check this for yourself, but the long-term average temperature of the climate of earth is from 15 to 20 degrees warmer than it is today. The only colder times were during "Ice Ages" (which we do not want), when the climate was only a few degrees cooler than it is today. Ocean level is rising a millimeter or so per year you say? Well, the "Sea Level" has only ever been a little lower than it is today. The average sea level across geologic time, and yes before humans discovered fire, is about 300 feet higher than it is today. The areas near our coasts, New York City and New Orleans and Miami are on average, under water. This is the main reason so many marine fossils are found in sedimentary beds well above the present sea level.

Commercial plant growers deliberately add additional $CO_2$ to the environment around their plants. Rates as high as 1,100 PPM are used, 3 times atmospheric level, which is only about one seventieth (1/70) of the level toxic for humans. There is also data that confirms that less, not more, water and fertilizer are required by plants grown in an atmosphere with enhanced $CO_2$.

Increasing $CO_2$ greatly increases the growth of vegetation. A greater amount of vegetation results in more "leaf area" facing the sun. Air temperature in a forest is much cooler than that above a bare field, or over a desert. For certain, with more plant leaf area, more moisture is pulled out from the ground and is transpired into the atmosphere. This transpiration causes cooling, greater amounts of clouds, and has a net cooling effect on Global Temperature. Higher $CO_2$ levels therefore cause lower Global Temperature.

With absolute certainty, more vegetation will supply more food for more people. If starvation of the multitudes increases, maybe we should burn much more coal and other fossil fuels, just to increase $CO_2$ level, so that the world is better equipped to feed its people.

Without question, if we were able to significantly reduce atmospheric $CO_2$, plants, wildlife, animals, and people would continue to starve to death, and in ever increasing numbers.

This leads to the conclusion that we MUST IMMEDIATELY either increase $CO_2$ in the atmosphere dramatically, or immediately halt and reverse the growth of the human population, and allow our forests and wild lands to recover. I pray to GOD that it is not war, disease and starvation which causes this, but rather responsible, sustainable, deliberate average proportional reductions in the rate of human fertility back to 2:1 (or less) offspring per mother in each society. Historically, no society has defeated any other by blatant, flagrant, uncontrolled reproduction. It just doesn't happen. This merely causes disease, starvation and social friction/violence in cities, amid a host of other tragedies.

$CO_2$ is absolutely necessary, and responsible both directly and indirectly, for virtually all life on Earth. An interesting, but significant change in the entertainment industry begs the question: Why have but few "$CO_2$'s rising, the earth is going to die" movies/documentaries been made since 2007? My guess is that the scientists on those special programs DO have a sense of ethics, and do not want to lie about the abrupt, downward change in temperature seen in 2008. There is very little financial interest now in supporting views other than the "$CO_2$ is bad" view. So, we're seeing repeats of the 2007 and prior environmental presentations and hearing more and more of the political decision that "$CO_2$ is bad". If there are 100 "$CO_2$ is bad" shows, and only one or two "$CO_2$ is good" ones, what will be the general public opinion? Does this make public opinion scientifically correct? Of course not.

The US's reputation for fixing real Environmental Problems is actually quite good. Recalled from recent memory, perhaps 30 odd years ago, there were wild screams of "Acid Rain". Rain is normally a bit on the acidic side, but globally rain had actually gotten noticeably more corrosive as evidenced on monuments and buildings. It was scientifically determined that sulfur dioxide emissions from coal and other fuels had a measurable effect on the pH of rain, so in the US we used practical methods of removing sulfur from coal and fuel oils. We have stopped adding lead to our gasoline, removing yet another real pollutant.

Sulfur has been largely removed as a U.S. emission, and this problem is for the most part remedied in most "Developed Countries". We do not, at least, hear much about the "Acid Rain" crisis anymore. Maybe, though, it was just another "media panic program" which was eventually explained logically, and never panned out.

How about fly ash or soot on glaciers? In the US, and in most developed countries, we have eliminated the majority of particulate emissions from Energy Use. If you've been around since the 1950's, or if you pay attention to old movies and documentaries, you know that factories, farms, transportation (tractors, trucks, trains, and autos) and individual home users of fossil fuels now emit virtually no particulates by comparison. Particulates are tiny particles of soot, of incompletely burned fuel, and are the things that make smoke actually look like smoke. Old movies show voluminous belching discharges of black smoke, even claiming this to be evidence of Human Progress, which in fact, it was. Picture in your mind a coal fired steam train plume of years past. Environmental films now use archive movie footage, or make videos of water vapor leaving factories to try to condemn Industry. These videos are shot during times of high humidity, so the "fog" water vapor leaving a facility cannot be seen to be mere water vapor, evaporating and disappearing once it leaves the immediate vicinity. Is this fair, or is it another case of "the ends justifying the means"? It's downright deceit, a product of devious intent. The next time you see an environmental film, pay attention to the white plumes of water vapor, and you will see nothing else, unless of course it's a 50-year-old film of a successful factory or power plant. The covers of this book show water fog emanating from factory exhausts and its subsequent evaporation. This is totally harmless in every respect.

People decided we needed to clean up our release of soot, of sulfur fumes, of PCB's, DDT, VCM's, other toxic chemicals, and we did. I hardly think that plant food, Carbon Dioxide, qualifies as anything bad, but rather as an ultimate good. As long as the Earth's concentration of Atmospheric $CO_2$ stays below the amount that exists on Mars (or 20 times what it is here today), everything will be fine. Due to the unlimited capability of plants to increase growth in response to more $CO_2$ in the atmosphere (for example during the Carboniferous Period), an increase of $CO_2$ in our Atmosphere to anywhere near 0.5% (more than ten times present levels) is virtually impossible. Humans simply do not have the capability to convert this much fossil fuel into $CO_2$, unless, of course, Human population increases 10-fold to 60 Billion. With certainty Wars, Disease and Hunger will not allow this degree of irresponsible, unsustainable growth to happen.

# MEASUREMENT, CAUSE & EFFECT, AND TRENDS

WASN'T THE WEATHER NICER LAST YEAR?

Many climatologists say our climate has been unusually mild for the most recent 1000 years. Only mild warm times and chilly times have occurred over this period. Before 1000 years ago, and during the times 10,000's and millions of years previously, the earth had seen multitudes of climate events of rapid change. The most recent 1000 years is characterized by both slow and gentle changes, including the questionable "global warming" or "climate change" we see today. Of course with our instant access to information, we know about climatic events as they occur. Before electricity and electronics and jet transport, we received far more of our information as anecdotal reports related by word of mouth.

As an example, the volcanic destruction of Pompeii occurred in AD 79, but it was not globally realized until AD 1748. Surely there were numbers of people in the Roman Empire who had heard about the Pompeii destruction, but globally it was unknown or a mere story, like the legend of Atlantis.

In the same way, the average person has very little reliable information about the long-term "average" climate, especially information regarding the numerous extremes experienced on Earth in the distant past. We actually have fairly accurate information about those extremes, but for some reason, most people do not think it is "natural" to have climate change. The truth is that Earth's climate is characterized as always being in a transition between extremes, and that in the present and recent past we have had a temporarily "good" (for humans) climate.

THERE'S A HOLE IN THE SKY

Another example is "The Ozone Hole" that we "found" over the Antarctic in the 1970's. Why did we find the Ozone Hole in the 1970's? Did the Ozone Hole form in the 1970's? Probably not. We finally had instruments sensitive enough to see the Ozone Hole. Many scientists speculate about the Ozone Hole, the dangers and causes of it, and money was spent, and continues to be spent, to study the Ozone hole. At one point, someone theorized that "CFC's" might act in the incredibly dilute chemistry of the upper atmosphere, to break down the UV protective Ozone. Therefore higher UV levels were allowed to pass down to the surface where they are bad for humans and other living things. Many CFC's have since been pretty much globally banned, but the Ozone Hole still exists.

http://web.archive.org/web/20070823125512/http://jwocky.gs
fc.nasa.gov/multi/aug-dec96.mpg

Here we can see nice computer graphics about how the Ozone Hole moves around and changes shape, and how it gets bigger and smaller over time. The gaping "hole" in the public understanding about the "Ozone Hole" is whether the Ozone Hole existed tens, hundreds, thousands, or millions of years before we could first measure it. We were led to believe that "humans did it", and that we must do something immediate and substantive "just in case". The Ozone Hole has likely been there all along, and just bounces around in size and comes and goes from time to time. But that's all history now.

MERCURY IN FISH

Another case is Mercury in Fish (Methylmercury). At one time Methylmercury was in industrial waste, the product of early technology that affected waste streams, but this is now corrected. Another source of mercury, for which hundreds of millions of dollars are being spent to reduce, is the miniscule amount of elementary mercury vapor released by combustion of coal. So, those in "forcing" positions are blaming all methylmercury in seafood and fish to this cause.

At some point in recent history, we developed instruments sensitive enough to allow measurement of methylmercury in fish. Pinnacle predators, such as swordfish and tuna have higher levels of this chemical than other fish, at levels around 0.90 to 1.0 ppm. We can now measure these levels, and we advise people to only eat certain amounts of these fish. This is great. Now, however, we are predominantly blaming our coal burning industries for this methylmercury. After all, we find methylmercury in fish, and we find that burning coal releases a minute but measurable amount of mercury. So, connect the dots ... Burning coal causes methylmercury in fish.

Unfortunate to this reasoning however, is the recent testing of "museum samples" of fish that shows methylmercury levels very similar to contemporary samples. So, there is no change in the amount of mercury in fish due to human activity, but the opportunity is being taken to condemn Productive Industry. The factor that distorts the human perception of mercury in fish is that there are tens of millions of samples collected today, and few precious museum samples, prompting us to ignore the "old and perhaps not reliable" evidence from the past.

Four hundred years ago, we could only detect mercury if there were visible droplets in a sample. Today, we can detect mercury at a level of 1 nano-gram in a liter of water. This is a sensitivity of 0.000001 ppm, or 1 part in a trillion (1 ppt). Since there are about 61,000 drops of water in one gallon, 1 ppt is the same ratio as one drop of water in 16,340,000 gallons of water. This is quite rare indeed. We could not measure mercury in fish just 100 years ago, but we can measure it today. The amount of mercury in fish has not changed, but our measurement has.

This has the same effect of sudden awareness created by improving instruments that led to the Ozone Hole panic. Maybe if the Ozone Hole doesn't close in the next 50 years, or once we determine that the Ozone Hole comes and goes as it pleases over thousands of years, regardless of what us humans are doing, we can go back to CFC refrigerant use. It cost humanity many Billions of dollars to change from using CFC refrigerants and propellants. Once we find that our politicians made a mistake, who will reimburse those thousands or millions of people who paid dearly for this change?

## HELP! MY HOME IS RADIOACTIVE - CALL A LAWYER

I'm sure that if many people suddenly had access to high energy particle detectors in their homes, and could measure the fantastic amount of sub-atomic particle bombardment we all experience on a daily basis, it would cause riotous societal panic. Can you imagine the overwhelming demand this would place on our elected officials? What would be the response of our media? How many resources would be spent for scientists to prove (not disprove, however) that this inhumanely cruel bombardment was from (think quick advocates, who has the deepest pockets?) some space program? Particle accelerators? Nuclear Power Plants? OH! It must be those damned compact fluorescent light bulbs! Maybe it's photovoltaic solar panels, or wind turbines! No, it must be from Hybrid Vehicles! After all, the general public was not aware of the massive bombardment of sub-atomic particles until after Hybrid Vehicles and Compact Fluorescents were available, now were they? Of course we know the constant bombardment of sub-atomic particles is a normal aspect of our environment, but the knowledge of same could be used to panic a part of society for a political purpose.

## WHAT'S REALLY GOING ON?

When we see something new, which we were not previously capable of seeing or measuring, there's always a danger that we will over-react to finding out about it. Our bodies are, for instance, about 90% bacteria, and the remaining 10% is human, by cell count. Makes one feel a little weird and concerned, doesn't it? It's perfectly normal, though, and not due to Industry, although before the Industrial Revolution, we did not know this, but now we do. So it is "logical" for people to think that a 90% bacterial "invasion" in every human is due to the Industrial Revolution, isn't it?

Today we sense and record each and every strike of lightning, every thunderstorm, every frost, and every accumulation of rainfall and snow. We document every flood of any significance, every tropical storm, every volcanic eruption, every "Sunspot", every minute deviation of temperature from average, and every new temperature record. So, the record shows that the present "Climate" has far more events than before "The Industrial Revolution".

So, does this convict fossil fuel use? Of course not. The past events are an incomplete record. It's not a valid comparison. A new high temperature record is set for Washington D.C. on a certain date. Does this mean it has never been that hot on that date before? No. It only means that since records were kept, it hasn't. So, it is likely that it has been much hotter on that date hundreds or thousands of years earlier. Of course about 4.5 Billion years earlier, the high temperature record on that date was probably in the neighborhood of 2,500 degrees. So, 90 degrees doesn't sound so hot by comparison.

Today's global communications have allowed a majority of people access to information about virtually every event as it happens. So, it seems as though there are now more and more serious storms and natural disasters than there were even 50 years ago. We're merely, however, only now more aware of these events.

Our real scientists are aware of, and consider the acceleration in detectability of physical phenomena. We can now, for instance, sample public sewage and measure how much, and which types of drugs, both legal and illegal, are being taken by the people upstream from the sampling point. Does this mean that no one used drugs before we could measure the levels in sewage? Of course not.

Fortunately, many accurate methods to measure things that happened "in the past" have been conceived and implemented. Tree ring growth is a standard method used to determine the rate of growth of trees in times before climatic measurements were recorded, and has been proven very accurate. Likewise, ice core samples from Polar areas are used to measure various atmospheric conditions back tens of thousands of years. Even sedimentary rock and sediments are accurately dated, and conditions in ancient times are accurately re-constructed through the chemical composition and types of fossils found in these sediments and rocks.

We have a preponderance of global temperature readings from the most recent 100 years, and low and behold, we can see that an almost imperceptible change has happened over this time period, and since people have been very busy over that 100 years, the anecdotal evidence suggests that people are responsible for the change. We can measure some small environmental chemical changes that have also happened in the same 100 years, and eureka! It must be humans. We must by now have billions of contemporary temperature readings, and the more time that goes by, the greater number of readings we have per second. So, we have a lot of confidence in this information. Other than a glaring mistake in a NASA report in which some Russian temperature data for August was repeated for September and October 2008, the data is indeed reliable, and trusted.

We also have a wealth of information about the earth's climate over the past thousands and millions of years, gleaned from geological "proxy" sources. These ancient samples tell us that:

On average, sea level is 300 feet higher than it is today. Sea Level today is low.

On average, global temperatures are 10+ degrees F higher than today. We do not want an Ice Age!

At present, we are actually within a few degrees of the conditions of an Ice Age.

Many paleoclimatologists agree that the most recent 1,000 years of climate have been unusually constant. Of course these mild times have existed numerous times in eons past.

Given what we know about the history of earth's climate, you can bet that it will get warmer, and the sea level will rise. These same predictions apply whether or not people live as they did in "the Stone Age", or in our modern lives, or even if there were no people here at all.

THE EGG OR THE CHICKEN?

The cause and effect relationship is core to most societies' legal systems. It is used in determining the Truth of who caused what to happen; the age-old adage of "chicken or egg, which was first?" We can use our logical minds, and say the chicken had to be first, to lay the egg, or that the egg was first, in order to grow up into a chicken.

Our legal system must depend on a time order of events, to determine the exact cause and effect, and the exact sequence of events is critically important in this process. In a robbery, for instance, we must establish the correct sequence and timing of events. If it is determined that a man with a limp was seen to be running as fast as he could from a store, and that it was later learned that a robbery had been committed in that store, was it the man with the limp who committed the robbery? We can speculate widely about the robbery, and come up with a lot of theories about it, but only after we find out all of the Truth, can we come up with a "Legal" Truth. If we learn that the limping man left the store before the robbery occurred, we must, of course, rule out that theory.

A large amount of information that we see today is presented in the form of graphs. Graphs facilitate our interpretation of information, particularly when dealing with data that is representative of a trend or pattern. When we see a graph whose slope is steadily inclining, it is natural to predict that this incline will continue into the future. However, there is no guarantee that this will be the case. The future is the future, and will always remain, to a large degree, unpredictable. If the future could be foreseen solely by historical trends, you would not be reading this, but would be at your stockbroker's office. One of the reasons that you're not at her office is that there is a magnitude of variables affecting stocks and bonds outcomes that neither you, nor anyone else, can be totally cognizant of.

The fact that a greyhound has won every race he has ever run does not conclusively prove that he will win his next race. Again, the magnitude of variables present precludes any assured foreknowledge. What the dog ate or didn't eat, the condition of the track, a shift in wind, crowd noise, etc. ad infinitum, can, and do, affect the outcome.

Graphs of trends can be misleading. For example, here is a graph of a climatic change occurring over time:

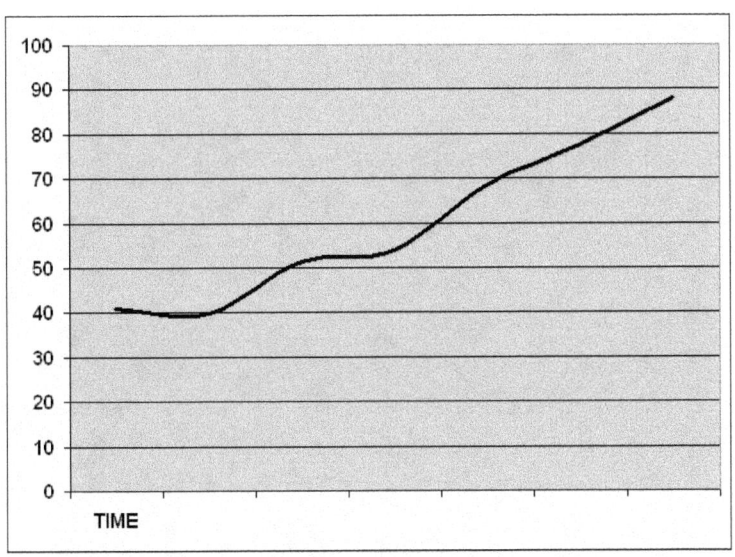

It would be reasonable to conclude that this increase would continue.

However, here is a larger portion of the same graph:

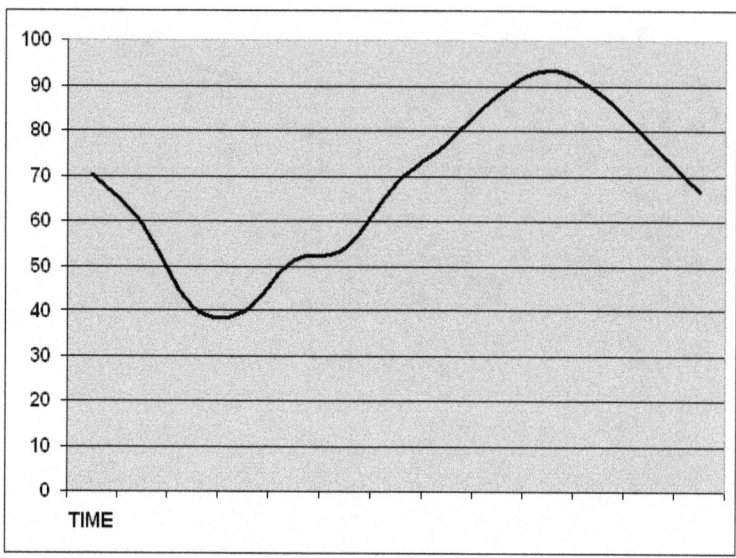

It is becoming clear that there may be a cycle involved. Or perhaps, the trend will continue to dive. The following shows a larger block of the same information. The pattern becomes clearer, and we would have an improved probability of predicting the future based upon the new knowledge of what we're looking at, and by knowing that monthly temperatures follow a reasonably predictable pattern.

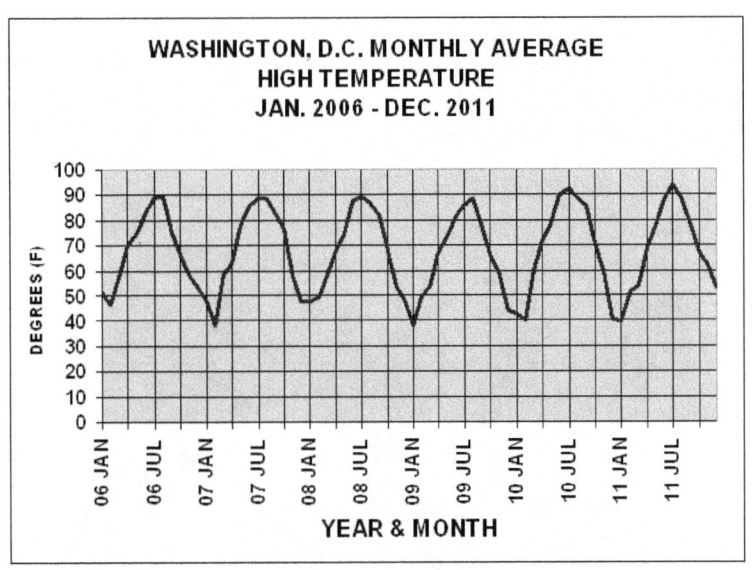

WASHINGTON, D.C. MONTHLY AVERAGE HIGH TEMPERATURE JAN. 2006 - DEC. 2011

If, however, we change the scale on a portion of the graph, and show increased detail, things start to look different:

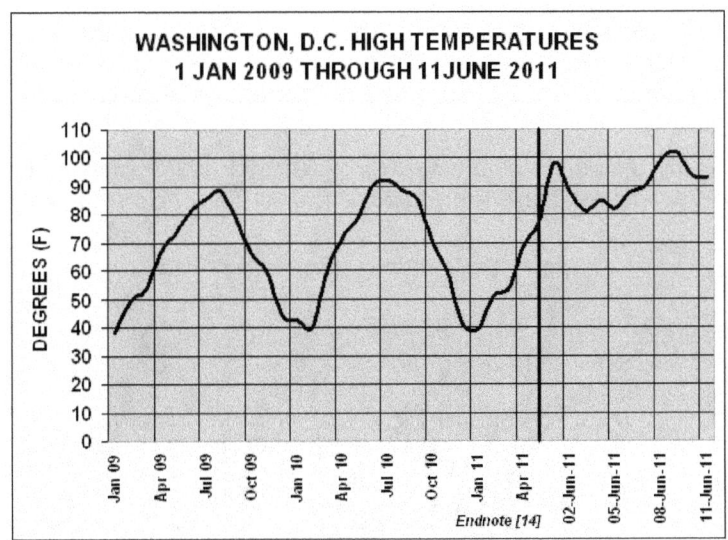

WASHINGTON, D.C. HIGH TEMPERATURES 1 JAN 2009 THROUGH 11 JUNE 2011

Without close scrutiny, we might think that things have somehow changed, and the previous cycles are no longer relevant. We might logically conclude that the recent climb in the chart is indicative of a new trend. This is not the case. This graph merely shows increased detail in June 2011 by "zooming" from monthly to daily data.

Here's a graph of cyclical climactic data:

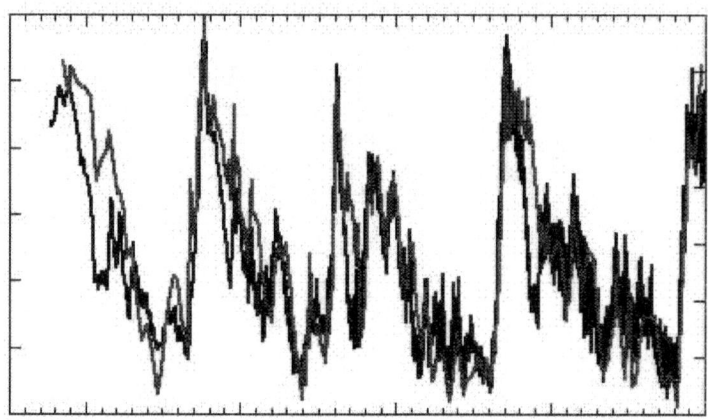

We could "zoom" in on the peak on the far right, and predict that the cycles are over, and that a consistent climb has begun, but is it logical to assume that the previous cycles are suddenly meaningless? The graph could go anywhere from its current peak, and we will only know the right answer after a passage of time.

My prediction of the future, based upon this graph, is that something like this is likely to occur:

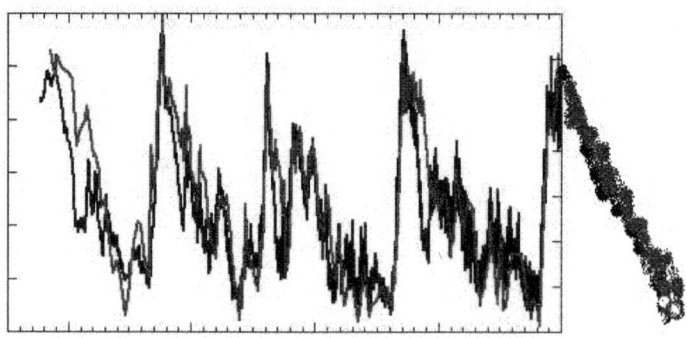

Do we have enough knowledge of all the variables to be

certain? No, of course not, but neither does anyone else. This graph appears again later in this book. I would hope that you remember this section when you see it.

We have a major problem today, with easy access to information (thanks Internet!), where anyone can get data, and interpret and publish it, without the benefit of "peer review". It is no wonder that the general population has become more skeptical about the information that is offered to them. Of course this is a time also of learning the fact that we must "consider the source". It will be those sources of information that become trusted which will be successful. The sources, which promote division of people by ethnicity, by income, by locality, or by religion, will fail. Those sources which promote panic and artificial emergency will fail.

## $CO_2$ - WHO NEEDS IT, ANYWAY?

Most of us, in our education, are exposed to a study of Chemistry. I recall two basic branches of that study. One is "Inorganic Chemistry", which looks at reactions of elements such as Hydrogen and Oxygen, which react to form water, or the reaction of Sodium and Chlorine, which react to form Salt.

These reactions seem simple by comparison to the other branch of chemistry, known as "Organic Chemistry". The distinction between these two branches is the addition of one chemical element to the organic side. That one chemical element, (in addition to Hydrogen and Oxygen), is the producer of life itself, Carbon. Of all of the 100-odd elements humans have recognized and named, Carbon is the element in our study of Chemistry that virtually defines the difference between "Organic" and "Inorganic". If there is no Carbon in a chemical compound, that compound is not likely to be considered to be "Organic". It defines the difference between the possibility of life and the impossibility of life. Several Science Fiction writers have proposed Silicon a possible alternative to Carbon for Organic Life. But, the vast majority of scientists associate Carbon inexorably with Life. Carbon is a part of ALL Life. Absence of Carbon is absence of Life.

I cannot emphasize enough the importance of the Big Four, the sole essentials for Life: The Sun for its heat and light, Oxygen in our Atmosphere, Water, and Carbon Dioxide. Pure Carbon, the element by itself, cannot sustain life. We can't eat charcoal and expect it to sustain us, although it is full of potential energy. We can't breath in Carbon Monoxide, a partially combusted product of Carbon and Oxygen, and indeed, we can't get sustenance from breathing in Carbon Dioxide. BUT, all the nutrition we consume, all starches, all amino acids, all vegetative matter, all proteins, all the mushrooms, eggs, meat, beans, and grains, absolutely EVERYTHING in the "Food Pyramid" is inexorably linked to, and is provided courtesy of the Provider of Life, Carbon Dioxide. We depend entirely on Photosynthesis, the conversion of $CO_2$, water, and Sunlight into food by plants.

If we could implement a technology to remove all of the $CO_2$ out of the atmosphere and therefore out of the waters (or vice versa), the assertions from enviropundits would lead us to believe that this would be the most wonderful thing that humans could do for the environment and the earth.

To extinguish virtually ALL Life on Earth, we need only to remove the Sun, or Water, or Oxygen from the Air, or Carbon

Dioxide from the Atmosphere and the Waters of the Earth. If Carbon Dioxide were removed from the Atmosphere, the "partial pressure" of Carbon Dioxide in the Waters would be higher than that in the Atmosphere, and $CO_2$ would then become depleted from the Waters, and ALL aquatic plants would die, along with ALL plant life on the surface of the Earth, and, in short order, ALL Animal Life on Earth.

So, no $CO_2$ in the atmosphere and in the oceans would be bad, really bad. Some think the ancient $CO_2$ level on Earth was as high as it is on Mars and Venus, until early life such as stromatolites started consuming it. Photosynthetic bacteria and early plants absorbed $CO_2$ and converted it into food, and released oxygen into the air, which was at the time toxic to most contemporary forms of life. Photosynthetic plants and bacteria were responsible for turning the atmospheric carbon into various deposits in the earth, such as crude oil and coal.

An increase in $CO_2$ may be the ONLY way we can supply the food for our fast growing population. Paradoxically, easier access to food reduces the amount of control a government can have "over" their people. Many, such as America's founders, believed "the people" should own and control the government and not the other way around - using food, debt, and ignorance to hold a society hostage.

A process to eliminate all $CO_2$ from the environment would be more destructive than even another mile+ sized comet strike such as the one that wiped out the dinosaurs, more destructive than another "super volcano" eruption in Yellowstone Park or in Siberia. Eliminating all the $CO_2$ would take Life on Earth back to the state in which it existed shortly after its formation.

THE CHICKEN & THE EGG, REVISITED, or "WE'VE BEEN GORED"

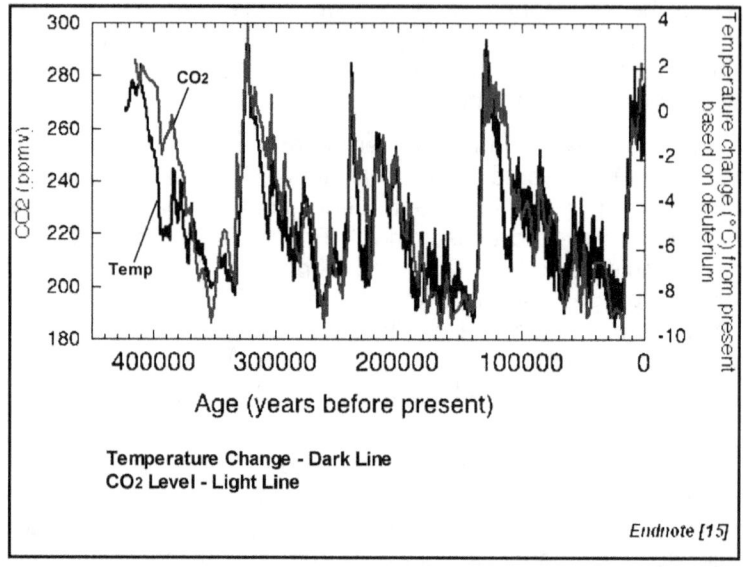

Temperature Change - Dark Line
CO2 Level - Light Line

*Endnote [15]*

This graph shows the obvious relationship between the Global Temperature and $CO_2$ concentration across 450,000 years. On a casual first inspection, we see the temperature and $CO_2$ levels change together, as in the famous graphs discussed in the movie "An Inconvenient Truth". Al Gore and his team use this data to "prove" that increasing $CO_2$ causes an increase in Global Temperature, hence "Global Warming". It's important for us to look at this graph with a bit more objectivity, however. On this graph, the dark line is the temperature graph. The light line is the $CO_2$ level.

Readily seen is the fact that AFTER temperature drops, there is a substantial delay and then the $CO_2$ level drops. On closer inspection yet, even when the temperature shoots up rapidly as it did about 130,000 years ago and again at about 15,000 years ago, the temperature went up first, then $CO_2$ level went up about the width of a line later. By separating the temperature and $CO_2$ level into separate graphs, as was done in "An Inconvenient Truth", it's really impossible to see that $CO_2$ changes follow temperature changes.

Perhaps a better title for Al's movie would be "A Convenient Mistake". Of course if one already believes that $CO_2$ level causes temperature change, Al Gore's graphs do nothing to convince them otherwise, nor are they intended to do so.

Indeed, the most striking point in "An Inconvenient Truth", is a Temperature/$CO_2$ graph which showed a consistent relationship between atmospheric $CO_2$ and global temperature, based on the paleoclimatological data. So, Al Gore and company do believe, trust, and rely on this data. Likely, they will also believe the pattern of the changing Sun/Earth heat relationship over that same time period, which gives a pattern of "heat applied to the earth" which rises and falls in a similar pattern to the $CO_2$ and temperature curves.

There IS proof that $CO_2$ level in the Atmosphere follows Global Average Temperature. The data used to make the graphs in Al Gore's movie actually prove this. I can't emphasize enough the importance of differentiating this relationship of "Cause and Effect". Lack of this differentiation was a significant failing of real Science in his movie. Let's put this cause and effect relationship in the correct context. The thing that happens first is the "cause". The thing that follows is the "effect".

The conclusion by Al Gore and his team is only one of several conclusions that could be made from that data, when we do not consider causality, as below:

1.) $CO_2$ goes up, therefore global temperature increases. *

2.) $CO_2$ goes up, therefore the Sun's heat output increases.

3.) Global Temperature goes up, therefore the Sun's heat output increases.

4.) Global Temperature goes up, therefore $CO_2$ increases.

5.) Sun's output goes up, therefore Global Temperature increases.

6.) Sun's output goes up, therefore Global Temperature increases and then (with some delay), $CO_2$ increases. **

7.) Sun's output goes up, then $CO_2$ increases, then Global Temperature increases.

8.) $CO_2$ goes up, then Global Temperature increases, then Solar output increases.

9.) Global Temperature goes up, then $CO_2$ increases, then Solar Output increases.

* = Al Gore's proposal

** = My rational choice

Since the graphs shown In "An Inconvenient Truth" are compressed to show 500,000 years of data on one screen, short time periods, for instance 100 years (one five thousandth of the width of the screen), cannot be seen. Studying the data more closely than in this movie, however, we can easily see that Global Temperature increases first, then after from decades to centuries, the $CO_2$ level in the atmosphere increases.

Again, the undeniable conclusion from the official US Government data is that temperature change is first, then $CO_2$ change follows. Therefore, temperature change causes a change in $CO_2$ in the Atmosphere. One theory about the delay in $CO_2$ change is that since the Oceans of the world are so vast, they warm up more slowly than does the Atmosphere. So, the air warms up, then decades to centuries later, the Oceans warm up. Warmer water cannot hold as much $CO_2$ as cold water, so the Oceans and Waters of the World give up a large amount of Carbon Dioxide to the Atmosphere. The oceans act as buffers, or "sponges" of $CO_2$. Heat them up, and $CO_2$ is "squeezed" out. Cool them down and the oceans "suck" the $CO_2$ back in.

AN ADDITIONAL MECHANISM

The amount of Carbon on Earth is basically fixed. We neither create nor destroy Carbon, and of course save for little additions riding in on the constant rain of cosmic dust, and the occasional comet and meteoric strikes, the level is basically unchanging.

We have carbon in the form of the most contemporarily popular compound, $CO_2$, in the air, and in the Waters of Earth. The remainder is wrapped up, at least temporarily, in that which we call life, and in Geology. Plants and animals hoard their share as do minerals including carbonates, hydrates, asphalts, hydrocarbons, lignite, bituminous, and anthracite deposits.

We have discovered there is a long-term relationship between the Average Global Temperature and $CO_2$ in our Atmosphere. Some (big Al) believe that more $CO_2$ causes higher temperatures, but as we can see this does not hold for planets such as Mars, previously found to have a concentration of molecules of $CO_2$ per cubic meter 20 times that of Earth, and it has no more global warming than on our Moon. So, we must accept the reality that $CO_2$ on Earth is only an indicator of the prevailing temperature. How can this be so? Once we consider the Natural relationship, it is quite a rational explanation.

Plant and animal life that has completed its purpose on this Earth is recycled into our environment. If the temperature is

144

"Warm", then leaves, animals, their waste products and the general "biomass" rapidly decomposes through Aerobic processes. Carbon is released from aerobic processes as Carbon Dioxide. It is recycled into the Atmosphere and Waters.

If the temperature is "Cold", however, the biomass does not immediately decompose, but settles down in bogs and low areas, and is covered by the relentless build-up of dust, silt, and additional layers of dead stuff. This material is not decomposed by the fast Aerobic processes, but is compacted and is slowly and only partially broken down by Anaerobic processes, forming some Methane, peat, and so on, with a large amount of Carbon incorporated into future "Fossil Fuel".

So, when it's HOT, the level of Carbon Dioxide in the Atmosphere increases, and when it is COLD, more Carbon is absorbed in the water as $CO_2$ (mentioned previously), and more Carbon is locked up in mineral deposits. Hence, high temperature causes an increase in atmospheric $CO_2$, and low temperature causes a reduction in $CO_2$.

A very nice aspect about the Carbon Cycle is that over time, there is no "Trigger Point" or "Tipping Point" or other "The Straw That Broke the Camel's Back" in the Carbon Cycle of our Climate. When it is Warm, and the level of $CO_2$ is higher (the Carboniferous Period, for instance), then much more plant life grows, cooling the environment.

BACK TO BASICS

Oxygen in our air is about 210,000 ppm and Nitrogen about 780,000 ppm, or 21% and 78%, respectively, as compared to $CO_2$ at 0.04% or 386 ppm.

The waters of the ocean and air always maintain on average the same balance of $CO_2$ relative to one another, after a period of equilibrium. If there's a higher concentration of $CO_2$ in the air, then the oceans absorb more. If the $CO_2$ in the atmosphere decreases, then the Oceans release more into the air. This balance primarily depends on the temperature and movement of the water in the Oceans. Cooler water holds higher volumes of $CO_2$. If ice melts and cools off the Ocean, then more $CO_2$ will be absorbed. Think about a carbonated beverage. When a glass of soda warms up, it loses its carbonation, and goes "flat". When the Oceans warm up, they release $CO_2$ into the atmosphere. Some estimate that at present there is about 50 times as much $CO_2$ dissolved in the waters of Earth as is in the air.

So this really makes sense. We get more heat from the Sun,

the Global Temperature increases, then after a delay of many years $CO_2$ in the atmosphere increases. Many Environmental Activists refuse to acknowledge this most logical conclusion for some reason, and seem almost desperate to implement industrial controls, which according to the above conclusion, won't make any difference at all. Are we supposed to allow ourselves to be economically crippled by this irresponsible action? By the "Carbon Credits" system? If Global Temperature is responding to OTHER planetary goings-on, it's counter-productive and economically crippling to try to discriminate against our society's Productive sector, which would be bad for the poor and bad for the middle class, due to skyrocketing prices for everything.

## CURRENT EVENTS

Global Average Temperature data from the 1970's through the magical year of 1998 showed a "probable" breakout from the 1000 years of stable climate. But then temperatures from 1998 through 2008 that showed predictions of future higher Global Temperatures were totally unfounded. If we begin with the Global Average Temperature from the previous ice age, about 20,000 years ago, we have indeed experienced Global Warming. If we start with the balmy temperatures of the Carboniferous Period, we indeed have on average, Global Cooling.

Here is a graph of recent Global Average Temperatures from data from the National Oak Ridge Laboratories.

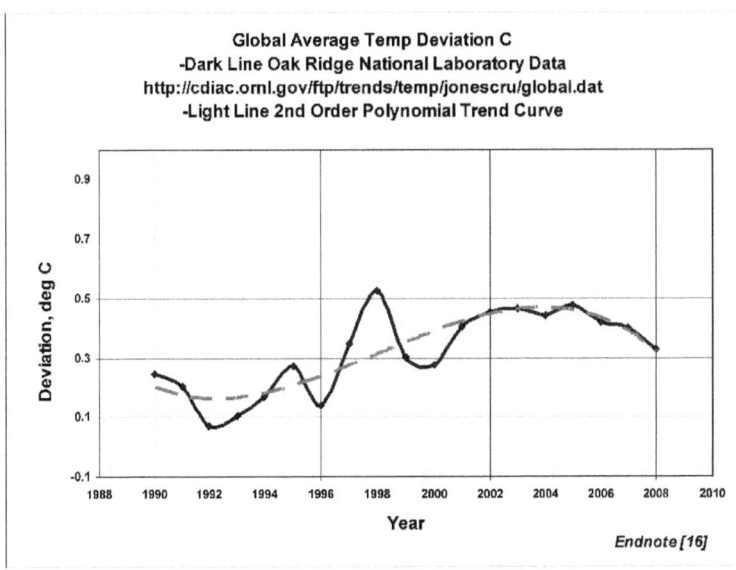

Global Average Temp Deviation C
-Dark Line Oak Ridge National Laboratory Data
http://cdiac.ornl.gov/ftp/trends/temp/jonescru/global.dat
-Light Line 2nd Order Polynomial Trend Curve

Endnote [16]

147

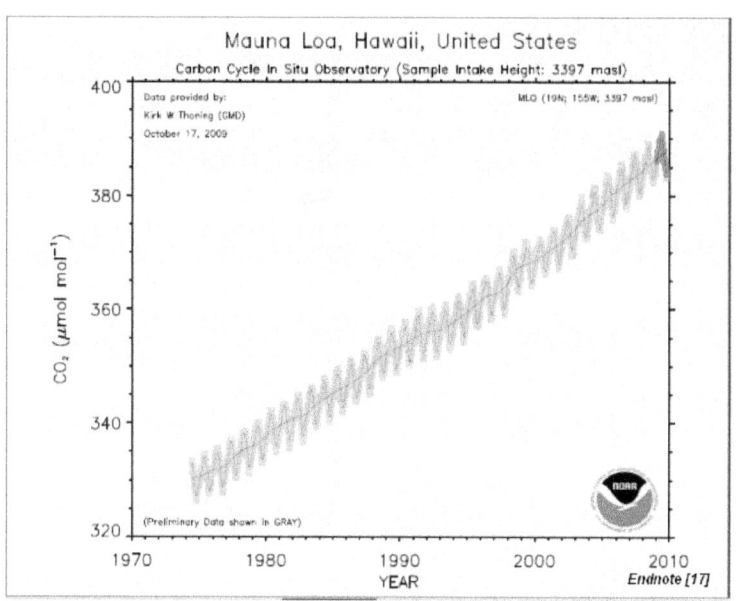

Endnote [17]

Atmospheric $CO_2$ levels have climbed from 365 ppm to 386 ppm during the time period of 1998 to 2009.

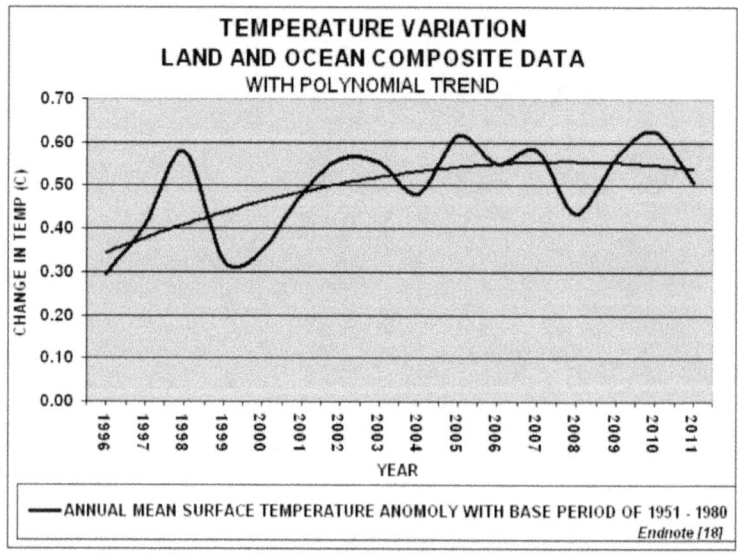

Endnote [18]

During this same time period, from 1998 to 2009, the Global Average Temperature slightly declined. Just as some scientists were trying to prove humans caused the cooling of the 1960's to

1970's, they are now trying to prove humans cause warming. Of course by switching to the more general term "Climate Change", any change, even just a change in the quantity of data available, will make it appear than human activities are responsible. Of course there WILL be Climate Change, as records during the human experience and in the geologic record prove.

Since $CO_2$ continued to increase into 2009, and the fact that 2008 was measurably cooler than 2007, $CO_2$ is NOT a significant (if any) cause of Global Temperature Change. Looking objectively and rationally at the scientific data, it's very difficult to not realize that <u>$CO_2$ change seen in proxy records is a RESULT of Global Temperature Change, NOT a cause of it</u>.

Since the "Global Warming" accusations are being found to be "faulty science" and losing social support, some squeaky little voices are starting to chant "Ocean Acidity". Yet another attempt to put something bad on those who use Fossil Fuels for the goodness for Mankind.

Let's talk a little about life and the carbon cycle, as this is a very relevant issue. Virtually everyone these days know that $CO_2$ in the air and water is absorbed by plants, and through photosynthesis converts water and carbon dioxide into food the plants can use, and into cellulose, the structure of plants. And, these wonderful plants release oxygen, which most people also know is necessary for animals to breathe, and is used to oxidize the carbon compounds in their bodies to generate energy. So, animals on land, in the air, and in the water release $CO_2$ back into the environment.

Farmers today are generating more productivity per acre planted. Of course some of this is due to smarter farming, to "no till", to improvements in machinery, improvements in natural and chemical pest control, and improvements in water and fertilizer use. An important part of this productivity increase is however due to increases in Atmospheric Carbon Dioxide, the fertilizer gas. Documented increases in plant growth prove a 50% increase in plant growth for a 100% increase in $CO_2$ concentration. So, if we start with 300 ppm of $CO_2$, and move to 386 ppm, that's an increase of concentration by (86ppm/300ppm), or 29%, we expect this is responsible for an approximately 15% increase in total plant growth across the entire Earth.

The natural response to higher $CO_2$ in the Atmosphere is an increase in the growth of plants. In such a rich Environment, some plants do not need to expend as much energy growing large leaves, which reduces the competition for sunlight, and there is then room for a greater number of plants to grow. For other plants, generation of biomass and growth of each plant and the fruit they produce increases significantly.

One relatively large-scale test on a forest in Tennessee using an increase in $CO_2$ level from 370 ppm to 550 ppm, an increase of 50%, caused an increase in plant growth of 24%. Commercial indoor growers often add $CO_2$ as a "fertilizer gas" at levels of up to 1,100 ppm which doubles plant growth and shortens growth time, while requiring virtually no increase in either water usage or necessary essential nutrients.

Greenhouses which are sealed and deprived of $CO_2$ from outside air have very stunted or halted growth as $CO_2$ levels drop to around 220 ppm. Slight leaks around outside walls help the plants around the edges of the greenhouse, and plants in the center show greatly stunted growth.

Photosynthetic life (Plants, Algae, etc.) have reduced the $CO_2$ level to about 300 ppm over hundreds of millions or billions of years. As the $CO_2$ level plummeted, less and less Plant growth was possible. Now that Humans are found to be responsible for a fraction of contemporary increases of the $CO_2$ level by recycling fossil carbon, Plant and Crop growth is increasing nicely, and this is certainly needed.

We know that $CO_2$ levels can naturally be as high as 1,700 ppm or higher, or a bit more than 4 times the level it is today, and even as high as 5,000 ppm, as it was during the "Carboniferous Period". We also know that $CO_2$ levels can be pushed as low as 200 ppm during an Ice Age. So, these are "natural" levels of $CO_2$. Maybe at some point, (maybe 5,000 ppm or 10,000 ppm) there could be a measurable effect of $CO_2$ as a "Greenhouse Gas", actually changing Global Temperature, but it is not seen to have this effect on Mars with the equivalent of 8,300 ppm. But then again, maybe the MASSIVE increase in plant growth of several hundred percent over the present yields would re-forest all the deserts and plains, and cool the Global Climate excessively.

The results of a study done by the Carbon Modeling Consortium, a joint effort of NOAA, Princeton University and Columbia University were published in October 1998. This team's data, collected 1988 - 1992, found that North America is behaving as a "Carbon Sink", an area in which carbon dioxide levels drop because carbon being absorbed by plants exceeds the amount being produced by all sources in North America.

They cited massive forest regrowth in areas previously cleared by farmers as a possible factor. This study was undertaken when gasoline was approximately $1.00 a gallon and before gasohol, CFL lamps, and the other conservation efforts we see today. This study was done before the current recession, a time when production levels were higher, and restraints were less.

Indeed, if we factored out the increase in absorption of $CO_2$ by the explosive growth of photosynthetic activity, the $CO_2$ levels today would be much higher, even without combustion of Fossil Fuels. Growth of plants follows and tempers Global $CO_2$ level changes, and is negative feedback on Global Temperature.

A way to look at how $CO_2$ has an effect on Global Temperatures is to look at the effects in addition to the "Greenhouse Gas" theory. We all know that a desert, with virtually no vegetation, is much hotter than a dense forest at the same latitude. So, more vegetation is better for more moderate temperatures. With no $CO_2$ there would be absolutely no vegetation, as all plants would literally starve to death and the Earth would become one huge desert. When people chop down forests and convert land into cities and wastelands, there is less vegetation. With less vegetation, the Global Climate gets hotter. We have very accurate measurements of how much hotter cities are than country areas, even after accounting for the heating effect of energy consumption. With higher levels of $CO_2$, however, more vegetation and more crops grow, hiding more bare ground, and cooling off the environment. As example, some people say that North Africa and the Mid-East were once lush garden spots of the world. But populations increased a lot, trees and bushes were chopped down, promoting deserts with high temperatures. Perhaps this is a mere anecdotal relationship, and perhaps not.

Another fine example of converting natural areas from vegetative to desert areas is the Amazon Rainforest. In the near future, due to "slash and burn" farming practice (which yields a short burst of lumber then one or two years of good farming), the Amazon Basin will become a massive desert. The loudest outcry against deforestation does not come from "Tree Huggers" actually, but rather from "Animal Huggers". Loss of habitat due to exploding populations in Third World countries is pushing some species towards extinction.

It is believed that without $CO_2$ levels of 1,700 ppm (0.17%), or well over 4 times the amount of $CO_2$ as in the air today, there would not have been enough green growth to foster the growth of Dinosaurs to such colossal dimensions. Dinosaurs would have been kept stunted to the size of today's mammals, say from the size of a cat to the size of an elephant.

As for higher temperature levels, if, after its initial formation and cooling some 4+ Billion years ago, the Global Average Temperature were to have stayed the same as it is during today's "1000 years of stability" some Billions of years later, perhaps the dominant species would be the Trilobite, the Nautilus, or maybe the Cockroach! Really big, arrogant, self-righteous cockroaches.

"Very recently, human effects have become evident, not yet showing both size and duration that exceed peak values of natural fluctuations further in the past, but with projections indicating that human influences could become anomalous in size and duration and, hence, in speed." [19]

The above quote from our government's main science centers states that evidence exists that humans have caused a change in the environment, without specifying what that change is or the degree of that change. At its most benign level, it's a slight, but measurable shift in the location and composition of naturally occurring chemicals, such as $CO_2$, lead, and other minerals. This quotation clearly states that those changes are NOT as big or as long lasting as natural fluctuations. Making "Projections" or trying to tell the future, yields a big MAYBE that humans' effects will be significant. Not a definitive yes, just a maybe, but then again, maybe not.

This most official of reports does not say, it only attempts to imply the statement: "Humans are responsible for Global Warming and/or Global Climate Change". If the data supported this conclusion, however, this report would say: "Humans are Responsible". But, it doesn't.

This in no way, of course, is meant to say that Humans have not added somewhat to the present level of $CO_2$ in our atmosphere, now just teetering on 386 ppm. This is however saying that human added $CO_2$ is not having a measurable effect on Global Temperature or Global Climate.

Humans are making a tiny change in the balance of the "Carbon Cycle". However, the Human effect is NOT causing an increase from the 300 ppm of "pre-Industrial times" to our current level of 386 ppm, but is in fact responsible for far, far less.

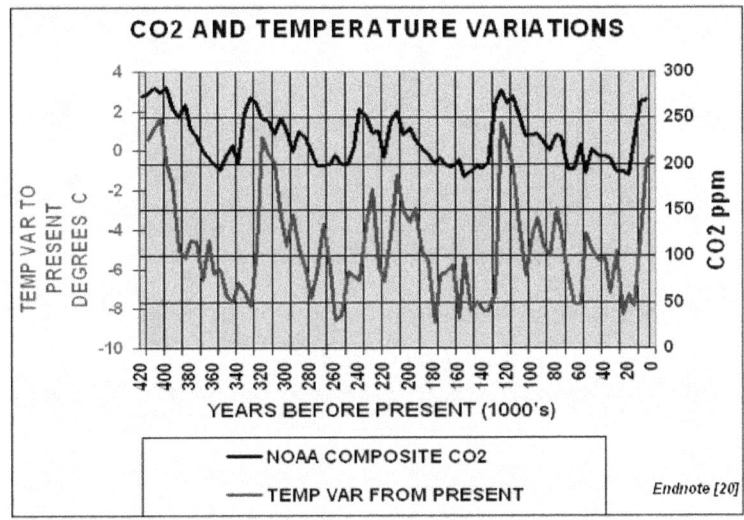

One piece of information we can recognize in this graph is that the Earth was on average much colder over the past 420,000 years than it is today. The Earth is more often in an "Ice Age" than it is at the nice, warm temperatures of the present over this time period. Most scientists agree that living in an Ice Age is a far more difficult Global Climate for people, animals, and plants than at the benign temperatures of today.

As previously stated, perhaps the biggest effect of the increase in $CO_2$ is some of the increase in crop yields we've seen, which heretofore have been solely attributed to improvements in fertilizers and farming technology.

The actual, real effect of "Human Activities" is an increase in the burning of "Fossil Fuels", perhaps at a rate faster than new Fossil Fuels are being formed, and they ARE of course being formed as this is written. As Atmospheric $CO_2$ level increases, and more plants grow, then the rate of formation of new Fossil Fuels increases. And as we entice oil and gas from beneath the surface of the Earth, the pressure in those deposits is reduced, and there is less natural seepage pollution by crude oil into our streams, rivers, and oceans, and less leakage of methane into

153

our atmosphere.

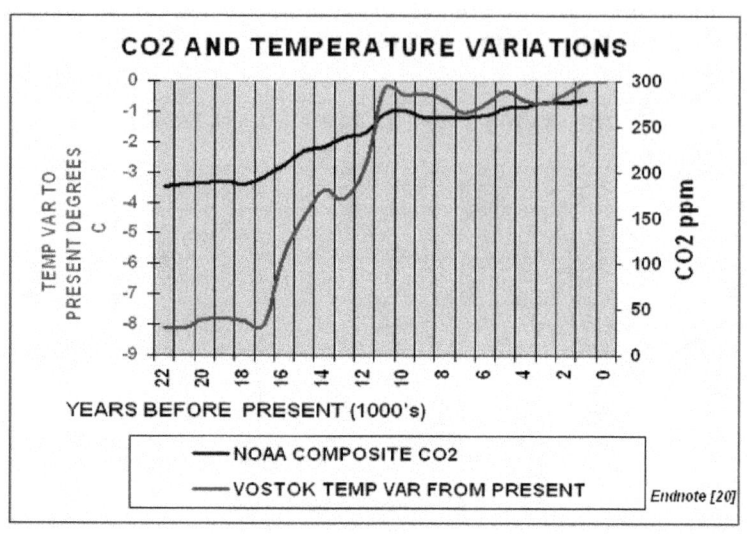

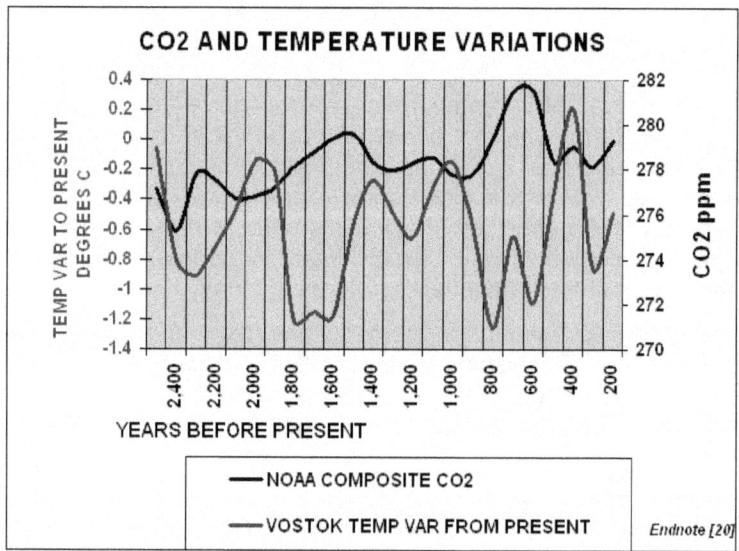

The above graphs show what our best scientists tell us about the record of global temperature from 22,000 years ago, until the present. The "Medieval Warm Period" is from the 9th to the 13th centuries AD (1,200 to 800 y.b.p.). The present temperature is at the end of the graph on the right side. There is certainly nothing obvious that indicates any abnormality about the present temperature, and nothing obvious or unique about the present temperature as compared to the temperatures even within the most recent 1000 years.

The egg/chicken question once again becomes paramount. Did the rise in temperature result from the onset of human civilization OR was the temperature change natural, which fostered humanity's growth?

Another undeniable conclusion from this graph is that temperature has changed far more in the past, even within the previous 15,000 years, then has changed since the start of "The Industrial Age" in about AD 1,800, now 200 years ago. "The Industrial Age" might be defined as the beginning of humans burning significant amounts of "fossil fuels".

In our discussion of Environment, we're really discussing the effects of humankind and Industry. True, we have had incidents that actually had a short to mid-term negative effect on things living in the Environment. Production of "Kepone®", a very effective, long lasting insecticide when used properly, was produced by an unthinking group of individuals, who dumped toxic waste chemicals in the James River. As a precaution, fishing industry downstream has been restricted in some areas for many years.

2008 and 2009 Global Temperatures have switched to a cooling trend, reversing the previous 35 warming years. There is evidence of another cooling from 1940~1970 (which by the way occurred during the greatest increase in fossil fuel usage in human history). In human history, we associate times of a cold climate with the greatest scourges of crop failures, food shortage, insect and rodent infestations and plagues (including the great bubonic plague). Warmer times are likewise associated with the greatest improvement in the human condition and the greatest surges in health improvement, population growth, and even technical creativity and the arts, such as during the extra warm times during "The Renaissance".

Even if we were to consider increases in atmospheric $CO_2$ to have a slight effect in the "earth blanket" warming effect, we need to put the effect in perspective with ALL "earth blanket" warming gases. Ignoring the effects of Nitrogen and Oxygen in air, water vapor (humidity, not fog or clouds) comprises 97% of the warming "earth blanket" or Greenhouse effect, and the total contribution of Greenhouse Effect to Global Temperature is not well known.

The conundrum of using Earth's changing atmospheric level of $CO_2$ and attempting to correlate it to Earth's climate through proxy data is the relative rarity of this gas. In a dried, filtered sample of Earth's air, the gaseous components are 99.96% <u>NOT</u> $CO_2$.

"The instrumental temperature record indicates that the Earth has warmed by 0.5C (0.9F) from 1860 to the present. However, this record is not long enough to determine if this warming should be expected under a naturally varying climate, or if it is unusual and perhaps due to human activities. Paleoclimatic proxy data can be used to extend climate records and provide a longer time frame (hundreds to tens of thousands of years) for evaluating the warming of the last 140 years." [21]

This quote, from the National Oceanographic and Atmospheric Administration, states unequivocally that the Earth has warmed by 0.5C over the previous 149 years. This is the actual, undeniable record of this change. Now most importantly, this sentence (underlined by this author):

"<u>However, this record is not long enough to determine if this warming should be expected under a naturally varying climate, or if it is unusual and perhaps due to human activities</u>."

There is nothing in any official, scientific pronouncement that states that human activities are, or are not the cause of climate change. There is an enormously increasing volume of data and confidence in our increasingly sensitive instruments, but even despite this, there is no actual link of human activities to "Global Warming", or the now popular term "Global Climate Change". All that is heard is pure speculation and unfounded assertions.

# HEAVENLY BODIES

In order to examine the possible effects of different variables on Earth's environment we can glean a large amount of information from what we now know about other bodies in our solar system. Since the dawn of the space age, we have accumulated massive amounts of data about Venus, Mars, and the Moon. How well does the current Global Change model being applied to Earth fit the facts if applied to these bodies?

Some definitions and data:

In Science a new term has come into recent use: "Albedo". In the most general case, Albedo is the amount of Solar Radiation (UV, Vis, IR, etc.) reflected by a Planetary Body, as compared to that incident on it in space. For example, if we had a perfect reflector which reflected 100% of all energy incident on its surface, we would denote this as having an Albedo = 1.00. The surface would not and could not heat up. If, on the other hand, it were perfectly absorbing, Albedo = 0.00, and if perfectly insulated on the shady side, the plate would get quite hot.

So, if 2 Planetary Bodies are the same distance from the sun, with the same Albedo, internal heating, 'shading', and thermal mass, then they will have the same surface temperatures. If one otherwise has less internal heating (less tidal frictional heating, and/or a smaller or non-existent nuclear furnace), then it would be colder. Manipulating only Albedo and leaving Internal heating and other thermal effects constant, the one with the higher Albedo will be colder, due to more reflection.

THE THERMAL FLYWHEEL EFFECT - SEPERATE FROM THE GREENHOUSE EFFECT

Those pushing their theory of a cataclysmic trigger point of a Greenhouse Effect totally ignore the major contribution of the Thermal Flywheel Effect in making our planet habitable. True, the Earth is 33 C warmer than a mass-less insulated ball in space, but we can see by less well understood thermal processes, that our atmosphere is not a 'hall of mirrors' to infrared heat. If it were so, Infrared Telescopes could not see through the atmosphere, but they can. There are actually only a few wavelength bands impeding infrared transmission through our Atmosphere.

The real challenge is to honestly overwhelm claims by "Junk Science" portrayals we see in our media, to separate out the Thermal Flywheel Effect to see the true, tiny fraction of warming that is attributable to the Greenhouse Effect, which is itself 97%

157

due to Water Vapor, with possibly detectable contributions from $CO_2$, methane and other gases. Of course clouds keep our daytime temperatures lower, and nighttime warmer, much in the same way as the Greenhouse Theory.

The Flywheel Effect is heat absorbed during the daytime by Air and Trees, Water, and the surface of the Earth, which is slowly released back out during the nighttime. A greatly simplified example, without using an Atmosphere, is a perfectly insulated disk in space, the same distance from the Sun as Earth, and one side is pointed towards the Sun. The front of the disk instantly gets hot enough to radiate away all the energy striking it, and the back side instantly cools to the temperature of deep space, 3 K, or -270 C. Using an imperfect black body, the front of the disk is 373 K, or 100 C. The average temperature is therefore -85 C.

If the disk was then changed to be perfectly heat conductive from the front to the back, it will have an effect the same as a Thermal Flywheel. In this case, half the heat is radiated away by the front, and half by the back, since they will both be the same temperature. Now, the amount of surface area radiating away the heat is exactly double. So, the temperature of the front will be cooler, and the back warmer. In fact, the front will be 314 K (41 C), and the back will be 314 K (41 C). The average temperature is then 41 C. Here the Thermal Flywheel has increased the average temperature by 126 C! Of course the Thermal Flywheel of the Earth is more complicated, but is separate from the Greenhouse Effect. As stated elsewhere, the Thermal Flywheel Effect is easily noticed by the delay in the highest temperature of the day from 'High Noon' to around 3 to 4 PM in the afternoon. The things that heat up relatively quickly for this day to day cycle are the low density things, like air, leaves, buildings, the top surface of the ground, and so on, that keep the temperature from rising quickly. Another easily recognizable Thermal Flywheel is the annual cycle. The longest daylight day of the year is June 21. The hottest day of the year is more like late July or early August, a month or more later in the year. This annual Thermal Flywheel is due to the heavy, dense things that take much longer to heat up, like bodies of water, rocks, and soil deeper from the surface.

The amount of daily Thermal Flywheel is thus dependent on the mass of the atmosphere and other 'low density' things, and also on the rotational speed. If a planet takes 1000 Earth years to make one revolution relative to the Sun, likely the hottest time of day will be very close to high noon (500 Earth years after midnight), and on the longest daylight 'day'. With a rotation speed of 24 hours, the Earth has far more of a Thermal Flywheel Effect, due our relatively high rotation speed.

My guess? I guess the Thermal Flywheel Effect is responsible for 32.5 C of the warming, and 0.5 C for the total Greenhouse Effect. Water Vapor contributes 0.485 C, leaving 0.015 C for $CO_2$, methane, and so on. Of the possibly 0.014 C due to $CO_2$, maybe 0.001 C is the total human effect, maybe less.

I am sure that a number of competent scientists are working on separating the "Elephant in the Room", the Thermal Flywheel Effect, from the 33 C that makes our planet habitable. They're doing this work in secret, to not fall under the knife of heresy. Science is Science, and "The Truth" is truth. Lies are lies no matter who speaks them. The problem with these Scientists publishing their work, is that it will be a total expose' on the whole 'Global Warming' or 'Climate Change by Humans' charade. Someone first announced their proof that the Earth is Not Flat, under risk of censure and imprisonment. Will you be one of the first to publish yours?

ALBEDO VS. ABSORPTIVITY

Albedo generally refers to the total reflectivity of a surface, measuring the reflectivity of heat (infrared), visible, and ultraviolet light. It is being said that visible light somehow magically turns into heat, but actually, it is a small part of the total energy. "Science" documentaries state that snow, with an Albedo of .90, prevents warming by reflecting solar heating back into space. Data from the University of California, Santa Barbara MODIS Emissivity Library however, lists the emissivity/absorptivity of snow to infrared energy (heat) to be a near perfect .95. This is almost identical to ice, water, and granite. So, although SNOW REFLECTS ALMOST ALL VISIBLE LIGHT, it is a near perfect ABSORBER of heat.

We can easily see the percentage of heat to visible light in sunlight through examination of our new friend, the LED bulb, when compared to our old friend, the incandescent bulb. 3 watts of LED light output gives off the same lumens as a 45-watt traditional bulb. The traditional bulb, like sunlight, produces a lot of heat relative to its visible light energy. In this example, the incandescent lamp produces 93% infrared heat energy.

Although not a scientific approach to dispelling the myth that snow reflects HEAT away from the Earth, might it be asked: "Why does freshly fallen pristine pure snow melt much more slowly in the shade of a tree or in the shadow of your house?" It is because SNOW ABSORBS THE VAST MAJORITY OF SOLAR HEAT ENERGY radiating down on it.

It takes a lot of heat to melt snow, to change it from solid to liquid (44 BTU's per pound), compared to heating water which takes only 1 BTU per 1deg. F increase per pound. The same amount of heat it takes to melt 32deg. F snow into 32deg. F water will heat 32deg. F water to 76deg. F! It is no wonder that some people think that snow and ice are magically cooling our planet by getting rid of the heat.

It is not intuitive that since snow reflects visible light so well, that it is a near perfect heat absorber. In the "rules" we make for ourselves based on other observations, we equate a dark surface as a heat absorber and a light surface a heat reflector. In the Laboratory we have what is called a "cold mirror", which reflects nearly all visible light for illuminating a specimen, but reflects almost no heat onto a specimen, which might otherwise destroy it. We assume, based on our own experiences, that a bright surface, like polished aluminum should reflect all heat, but snow and ice, like a 'cold mirror', are exceptions to "common knowledge", being almost perfect absorbers of heat.

Try this as a mind experiment. Let's say you have an infrared heat flood lamp, one of the 150 watt incandescent red ones for keeping hot dogs warmed, or for soothing stiff achy muscles. If one were to set up a lamp like this pointing downward at an angle, and shining onto a sheet of aluminum foil, the rays reflecting onto your face will feel quite warm. Now, if you replace the foil with a cookie sheet size tray of snow, your face is NOT warmed by the reflected visible rays, and the snow will start to melt rather rapidly. If this experiment were done outdoors at 30 below zero in a strong wind, the snow would melt little, if at all.

Repeating the same experiment with an incandescent "white light" flood lamp, both foil and snow will be blindingly bright, you'll feel a lot of heat on your face with the foil reflector, but no heat with the snow reflector, and the snow will melt. The same results will be obtained with a similarly sized tray of rough surfaced ice. Glass smooth clear ice will however reflect less visible light.

If this experiment were performed with an LED (light emitting diode) flood lamp, it would be blindingly bright with both foil and snow, but no heat on your face with either type reflector.

I recommend against performing this experiment, due to the

risks associated with bright lights in your eyes, and the electrocution hazards of electricity and water. A scientist may consider performing these experiments in a safe environment in the laboratory. DON'T TRY THIS EXPERIMENT AT HOME! REALLY! DON'T DO IT!

The point behind this passage is that some "Science Documentaries" totally ignore the reality of the affinity of Snow and Ice to nearly perfect absorption of infrared heat. Heat from the Sun is NOT bounced off snow and ice into space. Those shows really piss me off...Almost as much as ones that claim methane spontaneously ignites when it hits the atmosphere. That's simply ridiculous.

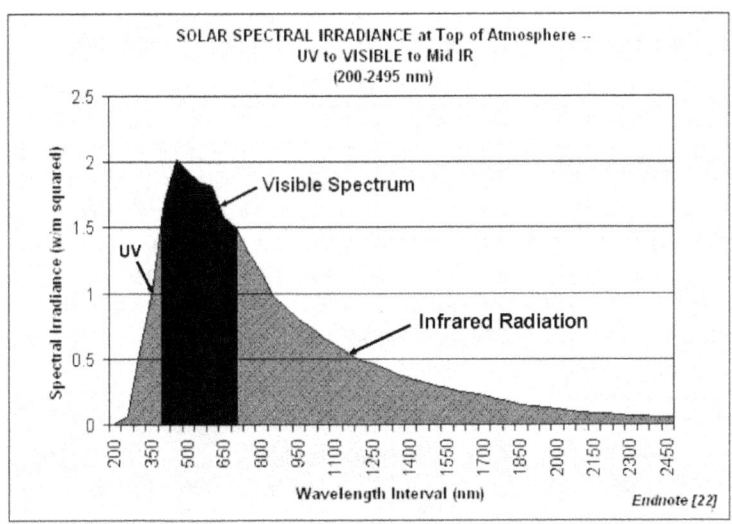

So, with visible light being a relatively small portion of energy in the Sun's spectrum as compared to the infrared heating portion (wavelength > 700nm), perhaps we should focus our attentions regarding heat balance on infrared radiation, recognized more than 200 years ago as being the heating part of electromagnetic radiation, using "emissivity" and "absorptivity" factors, rather than "Albedo" factors. It is reasonable to postulate that the Albedo approach to Global Temperature calculations is flawed, and that Infrared calculations are more accurate.

It is quite bothersome that 'Science' shows claim that Ice Ages spiral themselves into an ever colder cycle of cooling, the Sun's heat being bounced off into space. Ice and Snow beget more Ice and Snow, and wa-la! Snowball Earth. Those shows claim the Earth and Life is only saved by belching of an enormous cataclysm of Super Volcanoes releasing $CO_2$ into the Atmosphere. We do not have any proxy or recent evidence that an increase in Atmospheric $CO_2$ PRECEDED the end of the most recent Ice Age, 18,000 years ago, or for that matter, the end of any Ice Age. That Theory that would have us believe that more Ice cools the planet, and leads to more Ice is a one-way model, and if accurate would surely have lead to a terminal Global Disaster long before the present time.

We DO have evidence that massive volcanic discharges of dust into the Atmosphere have a net cooling effect on Global Temperature, similar to the now famous "Nuclear Winter" scenario. In this same vein, I propose that massive Volcanism is a more probable factor leading to Ice Ages, not ending them.

I, among a growing number of many others, am becoming more and more skeptical about "Human Caused" Global Warming or Global Climate Change as mere erroneous theories (like Cold Fusion), or perhaps as some sort of Conspiratorial Assertions. The evidence is simply not there. And now, we're hearing about more and more cases of deliberate falsehoods being spread by groups who should instead be practicing Real Science. Lying and Cheating to 'Win' has no part in Science, as it doesn't in Sports, and should not have in Politics.

THE DATA

Let's now look at Planetary Temperatures, and draw some conclusions about what the relative effects are contributing to their temperatures. Basically, there are only 2 sources of thermal energy, these are: 1) External source of heat from Solar Insolation, and 2) Internal Heating by nuclear fission and other processes.

Since Albedo includes ALL elements relating to Solar Insolation, and we have very accurate measurements of Albedo by Satellites, we can directly infer the contributions that are being made by Internal Heating.

| | ALBEDO | TEMPERATURE VARIATION FROM PREDICTED DEG. F | INTERNAL HEATING | ATM. $CO_2$, % | RATIO OF $CO_2$ MOLECULES PER VOLUME COMPARED TO EARTH |
|---|---|---|---|---|---|
| MERCURY | 0.07 | 0 | LITTLE | 0 | 0 : 1 |
| VENUS | 0.90 | 996 | VERT HIGH | 96 | 83,000 : 1 |
| EARTH | 0.31 | 61 | SOME | 0.04 | 1 : 1 |
| MOON | 0.11 | -32 | NONE | 0 | 0 : 1 |
| MARS | 0.25 | -4 | LITTLE | 95 | 21 : 1 |

TEMPERATURE VARIATION -
RESULT OF CALCULATIONS BASED UPON STEFAN-BOLTZMANN LAW
APPLIED TO ALBEDO, DISTANCE FROM SUN, PLANET SIZE,
BUT NEGLECTING THERMAL MASS (THERMAL FLYWHEEL) AND ROTATION SPEED

*Endnote [23]*

Regarding Internal Heating effect on Earth's Environment: The hottest location on Earth is either Death Valley, CA, or the Danakil Desert, Ethiopia. At both these locations, the Earth's crust is, for whatever reason, much thinner than average. So the surfaces of these places are much closer to the molten rock below and are hotter. Less distance for heat to travel = more heat transferred, and therefore higher temperature. The Earth's core is estimated to be 13,000 degrees F, cooled from the original formation of the Earth as much as it will over the next billions of years. There are vast amounts of high density decaying radioactive elements in and around Earth's core, providing the heat that preserves the 'fluid' environment necessary for maintaining Earth's protective magnetic blanket. This blanket in turn deflects the ion winds from the Sun, which stripped the atmospheres of Mercury, our Moon and Mars, whose interiors are now frozen solid through to the core, and have therefore insignificant protective magnetic fields.

The radius of the Earth is about 4,000 miles, the first 3,970 miles of which are metallic and molten rock. This molten rock terminates at a boundary between magma and our 30 mile thick crust. Temperature of the magma under our crust is in the ballpark of 2,000 F. The temperature falls at a rate of tens of degrees per mile from this point toward the surface. Crude oil temperature, as it is being presented to the surface, can be very hot, depending on the origin depth. The temperature of a mine just 2 miles deep is 140 F. Below this point, it is too hot for people to work even in short relays. Two miles deeper, and it's hotter than steam. On our planet we sit atop a huge pressure cooker, hence the interest in, and efficiencies of geothermal power.

If the heat from the interior of the Earth did not escape into our Environment and radiate off into space, there would be a gradual re-melting of the crust, and Life on Earth would not be a point of discussion, (to say nothing of the presence of this book.)

THE MOON

If our Moon had an atmosphere of roughly 15 pounds per square inch (Like Earth) the Moon's average temperature would be much closer to that of Earth, even without any water vapor or other 'greenhouse theory' gases.

Earth is much larger than our Moon. The Earth has an Atmosphere of predominantly Nitrogen, Oxygen, and Water Vapor, and our Moon has virtually no Atmosphere. The Moon is stone cold, solid through and through frozen rock, while the Earth is 97.8% molten rock with a nuclear furnace at its' core. So, with a large Thermal Flywheel and internal furnace, the Earth's surface temperature is about 30 deg. C warmer than the Moon's. No surprises here.

MARS

Although Mars has a lower pressure atmosphere than Earth, the high percentage of $CO_2$ molecules within its' atmosphere results in over 20 times the number of $CO_2$ molecules per cubic meter as on Earth. Mars has virtually identical "global warming" as our Moon, which has virtually no atmosphere, and no $CO_2$. This shows that higher $CO_2$ levels are NOT causing global warming on Mars.

This is further supported by the albedo predictions of the temperature for Mars, which shows this planet is only 4 degrees (F) from the expected temperature.

So, as long as we have less than 8,000 ppm of $CO_2$, (which would give us the same number of molecules of $CO_2$ per cubic meter as on Mars) we know for sure that we're OK. We now have around 400 ppm. Viewed in comparison to Mars, our amount of $CO_2$ is just not significant to global temperature or climate change.

VENUS

Venus is very hot, and it is theorized and promoted that $CO_2$ is playing an important role in Venus' high temperature, while others say that the length of a day on Venus, about one Earth-year per rotation, is causing this. But, there is 83,000 times as much $CO_2$ per cubic meter in Venus' atmosphere as compared to Earth's. Venus has, considering the atmospheric pressure is over 90 times that of the Earth, the equivalent of 32 million ppm of $CO_2$ if at one of our atmospheres. (1 million ppm is 100% pure.) Given the fact that the atmosphere of Venus is 95% pure compressed $CO_2$, comparison to a planet (Earth) with trace amounts of $CO_2$ at far lower pressures, is invalid. Venus' $CO_2$ data is literally 'off the chart'.

It has been said that Venus suffers from a "Runaway Greenhouse" effect. This implies that Venus was at one time cooler, and then through some trigger event heated up uncontrollably due to $CO_2$. I, among many, believe Venus to be merely cooling much more slowly from its formation than are other planetary bodies. The virtually opaque atmosphere of Venus may well be insulating that planet, restricting outward radiation. Venus' unique conditions virtually eliminate its sensible use as a standard for thermal comparison to other planetary bodies.

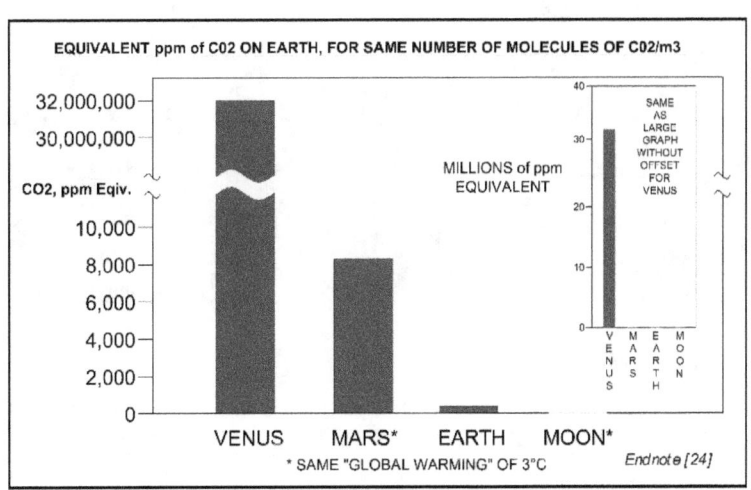

165

The most logical theory we can draw from these facts is that Venus has one hell-of-an internal nuclear furnace. In fact, radar imaging of the surface of Venus by NASA has shown Venus to have more evidence of volcanic activity than any other planet in the Solar System.

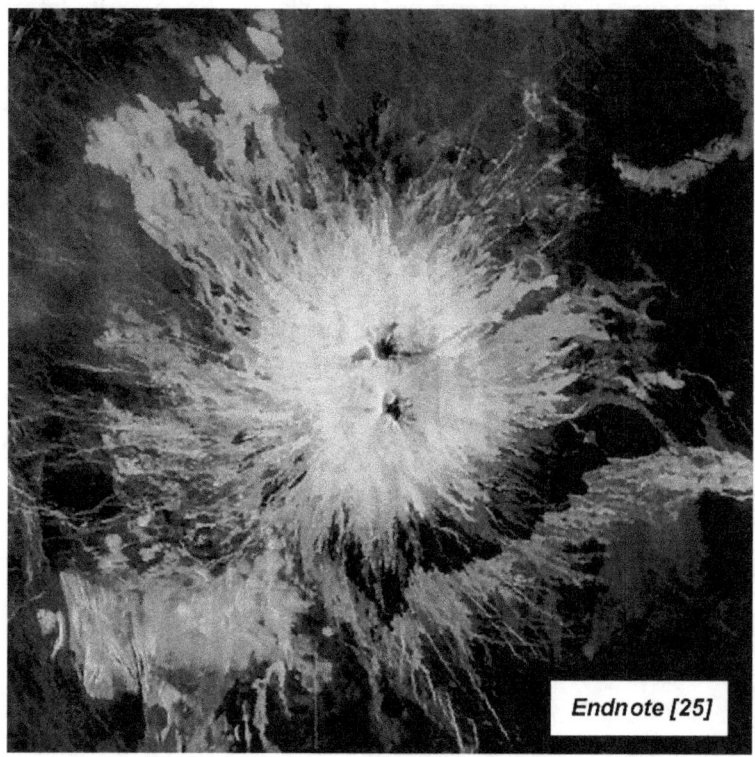

*Endnote [25]*

## HOW MUCH $CO_2$ IS TOO MUCH?

The concentration of $CO_2$ molecules per cubic meter for Venus and Mars, as compared to Earth, are in the ratios of 83,000:1 and 21:1, respectively.

We do not have scientific proof for the 'ideal' level of $CO_2$ on Earth. Earth currently has a $CO_2$ level of about 400 parts per million (0.04%). We know that too little $CO_2$ will not support photosynthesis on a scale adequate to feed the current food chain. There is a theory that $CO_2$ is a significant 'greenhouse gas'. We have proof that more $CO_2$ creates more green growth, which in turn reduces global temperature. We must be aware that If all land area on earth had no plants, the earth would be much hotter - One big desert surrounded by oceans.

We cannot scientifically state that 1000 ppm or 5000 ppm or 8000 ppm (Mars equivalent) $CO_2$ is too much. It is thought that the 32,000,000 ppm equivalent on Venus IS too much, but we do not have scientific evidence ruling out other causes for Venus' high temperatures to prove that 32,000,000 ppm is too much from a perspective of global warming.

We CAN say, with absolute certainty, however, that we know how much $CO_2$ is NOT too much. The amount of $CO_2$ on Mars, equivalent to 8,320 ppm on Earth, is NOT too much. Mars has very little 'global warming'. We can say with certainty, that for mammalian life, 70,000 ppm $CO_2$ is too much, not because of global warming, but due to its suffocating effect.

So, from a global warming perspective, 32,000,000 ppm of $CO_2$ might be too much, 8,300 ppm is NOT too much, and on Earth the current level is 386 ppm.

If we had 8,000 ppm (0.8%) $CO_2$ in our Atmosphere, many life-forms would probably evolve to new types, some would go extinct, and plant growth would be roughly 4 times the current level. Since growth of photosynthetic life absorbs $CO_2$ converting it into chemical energy, the cooling effect of increased biomass is a negative feedback on Temperature due to increased $CO_2$.

# THE EARTH IS SINKING AND THERE'S A HOLE IN THE SKY

## SEA LEVEL

We find fossils of sea animals well above the current sea level. Some are high in the mountains due to the land having been shoved up by extreme geological forces. But some are at these elevations due to former higher sea levels. Sea Level is presently; however, about 100 meters lower than the average over the geologic time scale. In terms of "Ice Age Cycles", we are still climbing up out of the last Ice Age. The present "Average Global Temperature" is a number of degrees lower than the long-term average highs. Since the Earth has been in a long-term warming trend over the past 15,000 years, even before people burned significant amounts of "fossil fuels", we should expect the temperature to continue to increase, regardless of what people are doing at present.

120,000 years ago, the sea level was about the same as it is today. However, starting about 120,000 years ago, it gradually declined over the millennia, and 20,000 years ago, was as low as 120 meters, that's almost 400 feet, lower than it is today. In the past 2,000 years through modern times, there continues an average sea level rise of 6 mm/year. Of course the rate of sea level rise has increased and decreased due to the freezing and melting of "permanent" ice on land, as Global Temperatures have gone up and down over this time period. Those changes in the amount of "permanent" ice reflect changes in Global Climate.

The long-term climate behaves much like our daily weather: "If you don't like the weather, just wait a bit". Since weather changes so much from day to day, and season to season, we can easily accept that there are undeniable natural causes, such as daytime vs. nighttime or annual seasonal changes. There are other Natural changes, which have, and will continue to make significant, extensive changes to our Climate. These Natural changes are of course devastating to some animals, to some people, and to some plants.

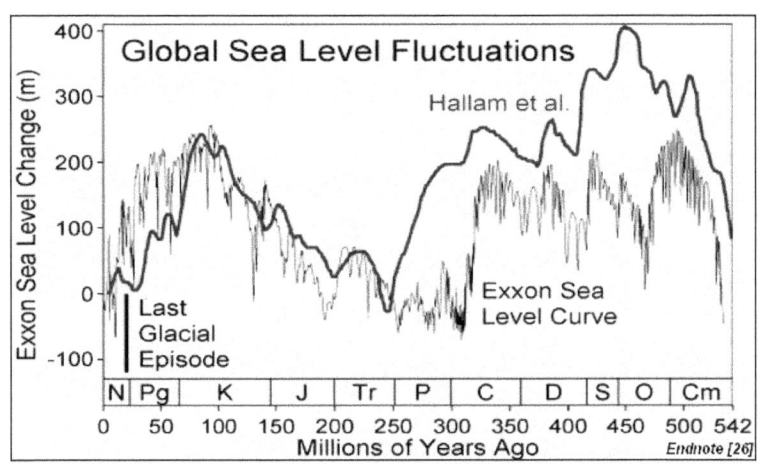

Endnote [26]

Paleoclimatic data show a direct relationship between $CO_2$ in the Atmosphere and the Sea Level. When the Sea Level has been 140 m lower than today, $CO_2$ level was only 200 ppm. In pre-industrial times, with about today's Sea Level, $CO_2$ level was 280 ppm. So, can we say that low Sea Level causes low $CO_2$? Or, is it a fact that since low Sea Level occurs during an Ice Age, when the Sea Temperature is also cold, that more $CO_2$ is absorbed in the colder water, so there is less $CO_2$ in the air?

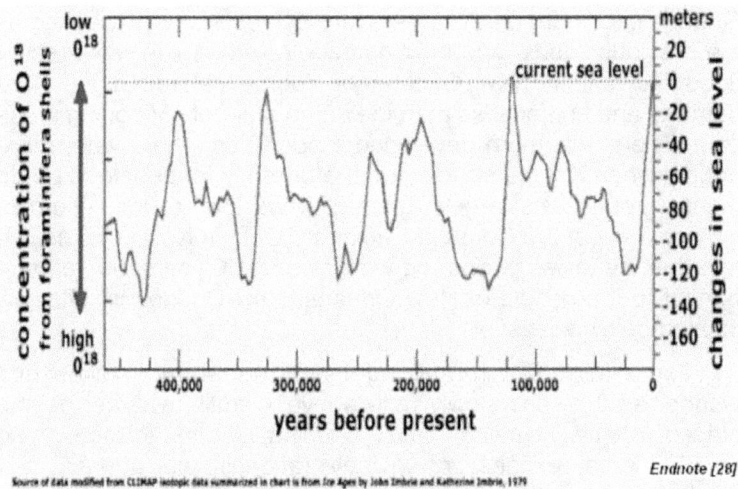

years before present

Endnote [28]

Source of data modified from CLIMAP isotopic data summarized in chart is from Ice Ages by John Imbrie and Katherine Imbrie, 1979

"The late Quaternary geologic record from various sources shows that sea level has fluctuated by more than 100 m over the past 20,000 years in response to climate warming and cooling caused by complex and largely unknown factors. In general, sea level has been rising, but at highly variable rates temporally and spatially." *[27]*

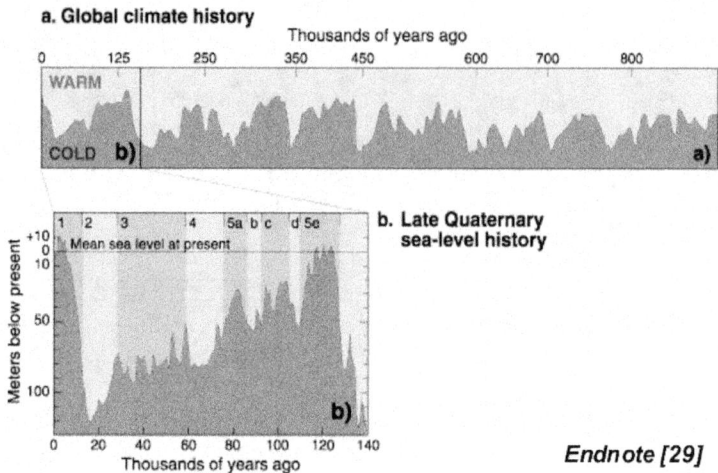

**a. Global climate history**

**b. Late Quaternary sea-level history**

*Endnote [29]*

KATRINA

The hurricane Katrina caused extensive damage in New Orleans. What was the cause? All too many are blaming the US Army Corps of Engineers, FEMA, and everyone else. New Orleans floods out on average every 50 years.... Who caused all the previous floods, and their damage? Katrina, the most recent flood incident in New Orleans of course caused a lot more damage and human loss then the same amount of flooding in the past. There was more developed property and there were more people when Katrina struck, and the urgency for people to get to higher ground was largely ignored. If we look at the record of performance of the Corps of Engineers work in New Orleans, we see that the levies performed well for over 50 years. 50 years of benefit to the people of New Orleans. New Orleans is, after all, located <u>below</u> sea level!

You won't catch me living below sea level. Makes one wonder as the seas slowly rose above Atlantis, whether people stayed there, cursing and blaming their advisors and environmental experts, or if they rationally considered their predicament, cut their losses, and moved on to higher ground?

Over the Geologic Record (BILLIONS of years) we know that sea level is on average more than 100 meters higher than it is today, so our civilization should be established on "high ground", or at least have plans in place to gradually move to higher ground, or build up the ground level as the sea gradually returns to its geologic average. Building Levees and pumping stations is too expensive, and is proven to be less than 100% reliable, but was a fun and challenging, albeit an expensive experiment. Building up the ground level, or "moving to higher ground" are much more sensible solutions.

You can bet that some people would blame a Sea Level Rise on those now living safely above Sea Level. Those above sea level are sensible folks not living in a Flood Plain, who are not living on the shore, or on a very low flat area near a coast. The coastal flatlands are flat, because they are on average, under water, and the silt carried down by rivers settles out on the bottom, filling in the low spots, making those areas flat. When sea level subsequently lowers, coastal flatlands are revealed.

Someone living comfortably above sea level may have a vacation cabin or a trailer in a low lying area, but they don't have all their eggs in that basket. How can we blame and tax these people, who are using good common sense, to pay for those who have not done so? Who would be the real victims? It's the old adage of the ant and the grasshopper. The grasshopper had a chance to do what it needed to do to plan for the future, and the ant did not ethically or morally owe anything to the grasshopper. The grasshopper did not want to, nor even offer to help the ant, much less to plan and carry out its own activities preparing for the inevitable winter. The ant could, at its own discretion, extend charity to the begging grasshopper. The Holy Bible tells us this charity would be good, but does not demand that it be given.

OTHER GOINGS ON

Other Climatic changes are common in nature. In the Western part of the US, for instance, there have been peaks of drought covering as much as 90% of the Western U.S., and at other times there was no drought, as has been reconstructed in graphs from 800 AD through today. Major shifts in the average pattern of drought in the Western US seem to happen about every 100 years, and there are fewer droughts in this area today (on average) than there were in the years between 800 and 1,300 AD, although there is a present trend towards more drought.

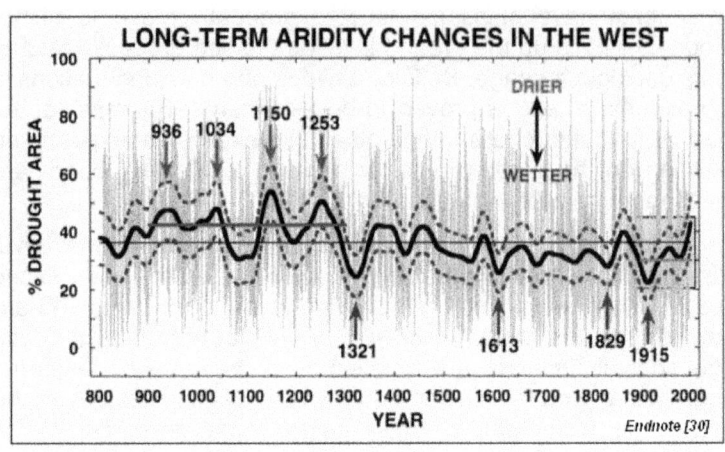

LONG-TERM ARIDITY CHANGES IN THE WEST

*Endnote [30]*

Our Global Climate has experienced "unusual" changes over the millennia. The one consistency in our Climate is in fact that it always changes, sometimes over long time scales, sometimes over short time scales, and this is all very natural. Temperature charts from written history and temperature and geological artifacts show there was a major Ice Age as recently as 18,000 years ago. Except for the "Medieval Warm Period", and an otherwise stable past few thousand years, our Earth has continued to climb up out of that Environmental Disaster of the most recent Ice Age.

Evidence in the Geologic Record, and some events recorded in the Human Record are proof that large and occasional huge chunks of debris have impacted the Earth. Many Scientists have compiled compelling cases that big impacts have been responsible for significant changes and indeed mass extinctions in the thread of life on Earth. Within the record of the Human Experience, we have seen extinctions of some species such as the Dodo bird, the Tasmanian Tiger, and the Ivory-billed Woodpecker. There is strong evidence that Humans have been responsible through their deliberate activities for a few of these extinctions. Some believe Humans hunted the Wooly Mammoth to extinction, others believe it succumbed to natural Climate Change. Fortunately for our American Bison, some rational forward-thinking people took action in time to prevent their extinction.

Today, scientists tell us we are living on a crust of frozen rock, which at a few 10's of miles thick, is proportionate in thickness to that of the shell of an egg. Below this thin crust is hot and molten rock and iron. If we took an average temperature of the earth, we would find about 99 percent of the Earth's mass is at least 1,000 degrees hotter than its surface. It's truly amazing that we have such a nice habitable region on the surface. It's even much more amazing that we people, responsible for far less than 100 ppm of the 400 ppm $CO_2$ in the Earth's Atmosphere, should have the gall to think we are having any effect on Global Climate, whatsoever.

OZONE

When did we first measure a hole in the Ozone? We first measured a hole in the Ozone when we developed the capability and confidence in that capability to measure it. We didn't know of the existence of a hole until we could measure it. Likewise, the then Planet Pluto was discovered in 1930 after our technology improved enough to detect it. It was a case of becoming able to detect Pluto, not that it was not there, and then it suddenly appeared.

"Ground based measurements of Ozone were first started in 1956, at Halley Bay, Antarctica. Satellite measurements of ozone started in the early 70's, but the first comprehensive worldwide measurements started in 1978 with the Nimbus-7 satellite. The Nimbus-7 data was rerun without the filter-program and evidence of the Ozone-hole was seen as far back as 1976." [31]

So, the British Antarctic Survey first found an Ozone Hole in the 1980's, and satellite measurements verified it existed in 1976. If we had satellites back in 1887, would we have measured the Ozone Hole then? My guess is Yes. You decide for yourself. Is it also possible that the "Ozone Hole" has come and gone over thousands or millions of years? Of course it's possible. Changing Climate and changing Atmospheric composition have continued for now Billions of years. It is also possible that since the Sun is too low on the horizon for much "UV-C" (very short wave ultraviolet light) to generate Ozone at the poles, the transport of ozone from areas closer to the equator is disrupted on a seasonal basis, creating the "Hole".

It is also possible that our Earth's magnetic field, which protects us from high-energy solar ions, is responsible for destroying some Ozone by directing those particles toward the Poles. This is, after all, the cause of the "Northern Lights" and "Southern Lights": High-energy particles being sucked into the Polar areas. If CFC's were really the cause of the 30% decrease in Ozone, one would logically say that since the majority of the CFC's were made and discharged in the Northern Hemisphere, the Ozone Hole should be over the North Pole. But, there is NO Ozone Hole at the North Pole.

We did not measure a hole in the Antarctic Ozone at the time of invention of CFC's. This seems an additional bit of evidence that Humans caused the Ozone Hole, now doesn't it? Of course we could say that it would take a number of years, and many tons of CFC's produced before an effect on the Ozone would happen. Yes, that is a reasonable and logical statement. The measuring instruments at that time were quite crude by today's standards, and not capable of measuring 0.3 ppm (0.00003%, or 1/3,000,000) of Ozone which normally occurs in the upper atmosphere, much less the difference between the normal 0.3 ppm and 0.2 ppm, the concentration defining a "Hole" in the Ozone. We claim we have a hole in the Ozone at levels 2/3 of normal (200 Daltons) or less. This happens once a year, for a period of a few months over the Antarctic. Interestingly, a movie shows the Ozone Hole from August until December 1996 on this Government site:

http://web.archive.org/web/20070823125512/http://jwocky.gs fc.nasa.gov/multi/aug-dec96.mpg

This video shows not only the Ozone "Hole", but also a ring with Ozone concentration much higher than normal (approximately 50% higher) circling around the "Hole". So, this shows strong evidence that the Ozone is not really being seasonally destroyed, but rather is seasonally displaced. The net result is the same: "The Ozone Hole". But, CFC's have been accused of destroying Ozone during a part of the year, and no research is evident of Scientists studying "Seasonal Displacement" of Ozone. "Seasonal Displacement" of course totally disproves the generally accepted theory of "Seasonal Destruction", the "Human Caused" political decision.

Seasonal displacement could be due to the Katabatic winds, the winds which rise up from the Tropics then converge and flow downward through the Ozone-containing air above the Antarctic continent on a seasonal basis. These winds can displace part of the 'Ozone Enriched' upper atmospheric air into the "donut ring" shape we see on the NASA animated movie referenced above.


HOW MUCH OZONE IS OK?

From 275 to 300 "Dobson Units" seems about normal over the Antarctic. If we concentrated this entire "Ozone Layer" to pure Ozone, it would be a layer over the whole Earth only 3 mm

(1/8 in) thick at sea level conditions. The Ozone layer concentration is about 0.3 ppm in a layer high in our Atmosphere, from 10 km to 35 km from the Earth's surface.

"Ozone is very rare in our atmosphere, averaging about three molecules of ozone for every 10 million air molecules." *[32]*

An interesting accusation was recently made on the news that studies show "sunburned" cells on the skin of our cetacean friends had been found. Immediately, this was attributed to the Ozone Hole. How can anyone possibly claim these measurements mean anything other than "some whales have sunburn"? Where is the data those accusers are using when implying that "Whales never had sunburn before"?

It is true also that some people living at or visiting beaches near the equator get sunburned. So, did the ozone hole cause their sunburn? Does Industry owe those people some type of compensation for their troubles? Or, is it possible that some people had sunburn before the 1940's? What are the results of studying whale skin for sunburn in museum samples from 'before the Ozone Hole? The silence of those previous accusers speaks volumes.

So although many people have accepted the "Scientific Theory" that Humans have caused "the Ozone Hole", many with a scientific bent are so far silently reevaluating the validity of this theory, as at this time, money is only granted to Scientists who will further a case for "Human Induced Environmental Damage". This strategy yields research dollars, and furthers a strategy for massively more government control of Industry and our Economy through various schemes of taxation and regulation. Where is the legal justice against the false accusers? Will Al Gore be sued by the Industries he has already cost countless Billions? What happened to fairness?

# CONCLUSION

## THE LIVES WE LEAD

We should all prize the concept of "equality of all people". Why then do some "look down" at others, trying to convince them that "their way" and their beliefs, and their actions are "better"? Are our Leaders and Media Moguls better than the rest of us?

Radically Left and Right people often believe "the ends justify the means". If all peoples' rights are stripped away so that there will be theoretical equality, will that be good? Do all people deserve the same things regardless of their Productive contributions? Should a group of people be discriminated by un-equal taxes? Should one person have to give 50% of their working life in taxes and another person virtually none?

When our elected representatives pass legislation that is contrary to the will of the people, they often lament the "tough decisions" they are being forced to make. This is another way of saying: "Who cares what they think, we'll do what we want". They do what they want, all right, despite the surprisingly accurate opinion polls not conducted by Special Interest Groups such as ACORN®, but by those polls conducted by long-standing objective groups. It is hard to believe that these "tough decisions" are always motivated by the best interest of constituents. We must ask ourselves what the real motivation is.

We can see Governmental domination and repressive, discriminatory taxation in a number of "advanced" societies today. It is left to The Will of those People to act to regain their Freedoms, their self-respect and self-determination.

Vilification (and "legal" process) of individuals, particularly individuals already in the public view, who speak outside "the accepted way" is a popular way to manipulate a society and rationalize all types of actions those in control are taking to implement their often hidden agendas. The case of Rod Blagojevich is a prime example of "sins against the party", rather than the case of just another corrupt politician. He became the "Fall Guy", distracting attention from other, more pertinent public concerns.

Throughout the history of the Western World, there have always been a few "crackpots". For example, a person wearing a "sandwich board" sign saying: "the world will end soon", saying that all books are evil, that the British will invade the Colonies, that the missions of our institutions must be expanded exponentially, or that they've seen both "Bessie" and

"Sasquatch".

Hitler was correct about the power of mass communications. We can influence large numbers of people to believe in a Perception. All that is needed is a strong emotional appeal to the irrational mind.

Recently, we have undergone an invasion by a virtual army, and have forgotten that our Constitution, including all of its amendments, is written for the benefit of U.S. Citizens. This includes the 14th amendment, which guarantees that babies born here of Citizens, and born of those here with valid visas, are in fact also Citizens.

Incidentally, The US Bureau of Land Management, the EPA, and the state of Arizona have teamed up to address the problem of trash discarded at our borders. They estimate that THOUSANDS OF TONS are dumped each year by those sneaking into our Country.

As the Government well knows, a 5% inflation in the value of money, (a dollar being worth a dollar to a dollar being worth 95 cents), brings in a tremendous income to the Federal Government. Let's only consider the estimated $5 trillion in people's "cash" savings. 5% inflation is 250 billion dollars extra Government income! If we consider also the investments, which of course lose value with inflation, then it's $1.25 Trillion extra income. This is totally immoral and unfair to those struggling to survive and to those planning on a self-sufficient retirement.

Counterfeiting is only very slightly different than government monetary policy which knowingly and deliberately creates inflation, no matter what the excuse. The difference between the two is that inflation benefits the government and counterfeiting benefits the counterfeiters. Can you say "Quantitative Easing"?

Should some people pay more for the same services? Isn't this discriminatory? Not only are higher earners paying more absolute dollars, they must contribute a larger percentage of their work year to society. This is truly an anti-incentive to performance, to excellence, and to Productivity.

The CEO's and the companies they guide are, for the most part, responsible for creating the massive increase in the average standard of living in our post-agrarian society. Is it the right of a government to create and control fear to gain compliance? Is it right, for instance, to control people's behaviors through intimidation and taxation?

Isn't it better to unshackle Industry and allow the massive improvements in Productivity to foster a Society to grow, which will naturally cause improvements in the "Condition of Man"? Was the relative "average standard of living" of people better before the Industrial Revolution?

Greed and Envy are economically destructive elements, are prime forces pulling down all of us, and are against the meanings in "The Ten Commandments".

If we do not consider the fact that the richest people create the most meaningful, Productive jobs,

If we do not consider that Productive employment opportunities and Productive Jobs add to the wealth of our Country,

If we listen only to proclamations that rich people have stolen from "The Poor",

If we only believe that new wealth cannot be manufactured,

If we only believe that some wrong in the distant past has "kept us back",

If we only believe that we are powerless to make any improvement in our own condition,

Then, are we willing to listen to the manipulative proclamation that "taxing the rich" (and their possessions when they die) will "fix everything"?

Sensible, rational, transitional improvements are required in the course of the way we do things. Revolutionary changes are only needed to rectify the misguided ideologies of the few. Biblically and morally, each of us has the societal obligation to support our families and ourselves. Fairness is NOT taking something from someone else. Taking something from someone else is theft, motivated by Greed. We do not all have to believe in the same goals and ideals. When someone else decides for us, particularly without respect for our opinions, then it is a bad way.

MAN AND THE ENVIRONMENT

So yes, we do have Climate Change. There will always be Climate Change. The Climate changed before Humans arrived and will continue long after we are gone.

BUT, our scientists have not reached a consensus on whether increasing the $CO_2$ level significantly will have more than a negligible effect, if any, on global temperature. Many of our politicians have decided, but they have done so without obtaining scientific consensus. The oppressive handle that a control such as "cap and trade", which many politicians are depending on, will cause the costs for all energy dependent industries to "necessarily skyrocket". People will have to pay enormously higher prices directly for energy, food, housing, and anything else that depends on energy in its manufacture. Aluminum, an energy intensive industry, will double or triple in price. Of course the poor cannot afford any increases whatsoever, so will we need to take even more money from our society's Producers and give it to the poor?

Evidenced by many accurate measurements of Weather, the Temperatures, Ocean Sea Level and Tree Ring growth, the natural, (non-Human Caused) variations in Climate of a magnitude that is a major impairment to humans and other life on Earth have repeated countless times over the millennia. These Climatic excursions happened in the times of the dinosaurs, the times of early humans, during recorded history, and of course continue to happen today. There is no evidence whatsoever, that any Climate Changes that have happened since the emergence of Mankind are any different in any way compared to the degree of Climate Change that happened prior to Humans. To the contrary, there were far greater Climate Changes before the arrival of man.

At present, the US is the global leader in total generated wind energy. The problems with wind energy are two: Adverse effect on wildlife (birds, bats, and insects - chop chop), and the reliability problem with the nature of the winds. An electrical grid requires a supply of power consistent with immediate needs. If winds suddenly decrease, the load must be quickly picked up by more traditional generating technology, and even gas turbines require too much time to pick up a load dropped by wind turbines.

We haven't heard about it yet, but it would not be difficult to scare many people into believing that using wind turbines decreases the winds, creates violent storms and extended droughts. It might be said their vibrations caused earthquakes and volcanic eruptions, and that whole classes of animals are being bludgeoned into extinction. If the wind generated energy industry becomes very successful, it goes without question that those "deep pockets" will be attacked, regulated, controlled and milked "for the good of the people, the plants, and the animals." And let's not forget how important this new source of Revenue will be in "Helping the Children".

When we burn coal with the correct amount of air, the exhaust is a clear, warm gas stream. And of course coal is not pure carbon.

We have found that burning coal releases sulfur when it is burned. Some very innovative people have figured out how to remove sulfur from the exhaust stream and actually use it as a feedstock for making certain products. Similar improvements have been made for liquid fuels used in automobiles and trucks, heating systems and for Industrial use, eliminating a source of "Acid Rain", which has been found to be unhealthy for the Environment.

Smog, now mostly an entry in history books, was a ground level reaction of pollutants, moisture and ozone, which, similar to clouds and dust in the air, reduced global temperatures. Changes made by Heavy Industries and Automotive manufacturers have largely eliminated un-burned hydrocarbons, certain oxides of nitrogen and other pollutants, which previously caused Smog.

Without a whole lot of effort, we could certainly convince an already emotionally charged crowd that pure "hydrogen hydroxide" is now being emitted in totally unacceptable amounts by industry, and that we must tax those anti-social groups mercilessly. Hydrogen hydroxide? Same as Dihydrogen oxide. $H_2O$. Water: As necessary for life on earth as is $CO_2$ in our atmosphere and in the waters of Earth.

The questions I will always ask myself when presented with a "new crisis" which seemingly demands immediate political, monetary or social action is "Does anyone stand to profit from this "new crisis"? Is there any real, verifiable information that supports belief that there is a crisis? Why am I hearing about this now? Unfortunately, I believe that crisis mongering is nothing more than just another emotional scam utilized by those in power to exert their will upon the citizenry.

POLITICS OF SCIENCE

What is the political affiliation of our Scientists? Could this have any potential bearing on their objectivity? Let's see...hummm. At one time, likely not. Only 6% of Scientists today claim they follow the most well known Conservative Party, the Republican Party. Most are liberals or progressives, known for their activities promoting greater governmental control over society, and the philosophy of "Economic Justice" or "Social Justice". How do you control Society? Control money. Control

180

Industry. Control Energy. Control Medicine. Control Food. Control housing. Get more and more people dependent on Entitlements, and hold them hostage, as we see happening today. This is the bondage of Welfare.

Throughout human history, sacrifices of people, animals, and valuables were made in appeasement to various supernatural deities, to try to control nature, or to defeat their enemies. Somehow or another in the US, "Carbon Credits" is poised to do the same thing, to make sacrifices of our industries, without any compelling real science. It is an emotional campaign. It is Chicken Little running around screaming "the sky is falling."

The term "Mission Creep" can be rightfully applied to an action of a "rights group" expanding their scope, particularly in the view that their original mission may become completed. They just don't want to go out of business or out of favor. For instance, when cancer is cured, what will be the mission of the American Cancer Society®? Will they become advocates for malpractice liability for today's cancer treatments?

One of the hallmarks of "The Scientific Method" is the acknowledgement of the critical importance of "Peer Review", which makes certain that good science is indeed good and truthful science. This review ensures that the science is not warped and distorted, and that the sanctity of Science is not manipulated by non-scientific political or religious external forces. A few examples of the value of Peer Review are the disproof of "Cold Fusion" and "Polywater."

How can a somewhat loosely associated group of organizations maintain their purpose when their central focus, "Global Warming" has disappeared, unless their "Mission Statement" is changed? What will be the use of trying to tax fossil fuel users based on "$CO_2$ causes Global Warming", if in fact $CO_2$ levels continue to go up, and Global Temperature continues to go down, proving $CO_2$ is not a significant cause of Global Temperature increase?

So true Science is always open to voices of reason and to review by other scientists. Unfortunately, a Political decision about the state of Global Climate Change has been made, circumventing Peer Review on the subject, to the point that even questioning this "science" has already cost careers. THIS IS NOT TRUE SCIENCE.

So what was it like? Everyone was saying "The Earth is Flat". There was no scientific fact, only the chant of the ages-

"Just look, it's flat." Then a voice rang out, at the hazard of the life of the speaker. Then not only was the Earth round, it wasn't the center of the Universe.

Just now a few souls cry out in the dark. - "What about Mars?" The warming chant, the unsubstantiated conviction that the creators of value are to be taxed. It's all about the money. It's all about the victims of those governments not allowing property rights, those enslaving their own people to a life of ignorance, a life of servitude, a life without justice.

THE WAY AHEAD

One of our forefathers, Abraham Lincoln, said at the conclusion of his Gettysburg Address: "...that this nation, under GOD, shall have a new birth of freedom-and that government: of the people, by the people, for the people, shall not perish from the earth."

This brings us to much larger questions concerning the moral root cause of most social calamities in human history. The crux of the issue of human rights that needs to be addressed by any society is contained within the answers to the questions: Should people have the right to think for themselves? Are they too stupid? Are some people better than others? Are the "academically elite" better people than someone like Albert Einstein, who flunked out of school? Are our politicians better than the loyal working and proud "blue collar worker"? Are CEO's, the most visible representation of our Industries, villains?

Governments can and do lose track of their ultimate mission, and morph their duty and responsibilities in a direction of the Government's benefit, particularly to those in power in the Government. National or Federal Governments are, after all, monopolies, and they require ongoing restraint by the governed - those states and individuals of the nation.

"Government is not reason; it is not eloquence. It is force. And force, like fire, is a dangerous servant and a fearful master." - George Washington

Ultimately, neither I, nor anyone else, can tell you the correct thing(s) to do, to create a better tomorrow, to encourage making more things, to improve the average standard of living for all. I can say that the important aspect of standard of living is the actual improvement for the poorest of us, NOT the difference between the poorest and the richest. We find that the greatest differences in wealth actually exist in countries embracing Communism and Socialism, and in those run by dictators.

Satisfaction in the results of hard work, the joy in happy family matters, the respect for Productivity of others, the confidence in controlling one's own destiny, and the reward of giving meaningful charity, these are the things we were created to enjoy. Do not allow yourself to be manipulated by hate, fear, greed, and jealously.

Use your good sense. Think for yourself. Always remember that emotions can be instant and overwhelming, and that it takes time to "use common sense", to think about something rationally, and to be "level-headed" before making a decision or taking an action.

Our Country's future is what this is all about. "What do we do now?" is the real question. "What we do now" is the answer. It all depends on where we want to go. When "The Cause" is more important than The Truth...Then This is the root of most problems...

Good Luck, be fair to yourself, and to all others.

# ENDNOTES

1. http://data.un.org/Data.aspx?d=MDG&f=seriesRowID%
3a566


2. 14 TH AMENDMENT

Section 1. All persons born or naturalized in the United States, and subject to the jurisdiction thereof, are citizens of the United States and of the State wherein they reside. No State shall make or enforce any law which shall abridge the privileges or immunities of citizens of the United States; nor shall any State deprive any person of life, liberty, or property, without due process of law; nor deny to any person within its jurisdiction the equal protection of the laws.

Section 2. Representatives shall be apportioned among the several States according to their respective numbers, counting the whole number of persons in each State, excluding Indians not taxed. But when the right to vote at any election for the choice of electors for President and Vice President of the United States, Representatives in Congress, the Executive and Judicial officers of a State, or the members of the Legislature thereof, is denied to any of the male inhabitants of such State, being twenty-one years of age, and citizens of the United States, or in any way abridged, except for participation in rebellion, or other crime, the basis of representation therein shall be reduced in the proportion which the number of such male citizens shall bear to the whole number of male citizens twenty-one years of age in such State.

Section 3. No person shall be a Senator or Representative in Congress, or elector of President and Vice President, or hold any office, civil or military, under the United States, or under any State, who, having previously taken an oath, as a member of Congress, or as an officer of the United States, or as a member of any State legislature, or as an executive or judicial officer of any State, to support the Constitution of the United States, shall have engaged in insurrection or rebellion against the same, or given aid or comfort to the enemies thereof. But Congress may, by a vote of two-thirds of each House, remove such disability.

Section 4. The validity of the public debt of the United States, authorized by law, including debts incurred for payment of pensions and bounties for services in suppressing insurrection or rebellion, shall not be questioned. But neither the United States nor any State shall assume or pay any debt or obligation incurred in aid of insurrection or rebellion against the United States, or any claim for the loss or emancipation of any slave; but all such debts, obligations and claims shall be held illegal and void.

Section 5. The Congress shall have power to enforce, by appropriate legislation, the provisions of this article.


3. U.S. Census Bureau, Census of Population, 1850 to 2000, and the American Community Survey, 2010.


4. ftp://ftp.bls.gov/pub/special.requests/cpi/cpiai.txt

http://www.bls.gov/data/inflation_calculator.htm


5. https://www.cia.gov/library/publications/the-world-factbook/fields/2195.html

Note: because China's exchange rate is determine by fiat, rather than by market forces, the official exchange rate measure of GDP is not an accurate measure of China's output; GDP at the official exchange rate substantially understates the actual level of China's output vis-à-vis the rest of the world; in China's situation, GDP at purchasing power parity provides the best measure for comparing output across countries (2012 est.)


6. Federal Reserve Bulletin | June 2012

www.federalreserve.gov


7. http://www.treasurydirect.gov/govt/reports/pd/histdebt/

histdebt_histo5.htm

http://www.treasurydirect.gov/govt/reports/pd/histdebt/histdebt.htm

http://www.bea.gov/national/index.htm#gdp


8. NOTE: *The date refers to the Department of War. The Department of Defense was officially created in 1949: The Department of War (1789), the Department of the Navy (1798), the Department of the Army (1947), and the Department of the Air Force (1947) were all reorganized under the Department of

Defense in 1949. see www.dod.gov

†Cabinet-level rank under George W. Bush.

www.whitehouse.gov/government/cabinet.html

ADDITIONAL SOURCE: Cabinet Department websites.


9. http://www.bea.gov/iTable/index_nipa.cfm

Table 1.15 Gross Domestic Product

Table 3.2 Federal Government Current Receipts and Expenditures


10. http://data.bls.gov/timeseries/LNS12300000

Employment population ratio (employed over age 16 on June1 of year shown as percentage of population)


11. 17th AMENDMENT

Clause 1. The Senate of the United States shall be composed of two Senators from each State, elected by the people thereof, for six years; and each Senator shall have one vote. The electors in each State shall have the qualifications requisite for electors of the most numerous branch of the State legislatures.

Clause 2. When vacancies happen in the representation of any State in the Senate, the executive authority of each State shall issue writs of election to fill such vacancies: Provided that the legislature of any State may empower the executive thereof to make temporary appointments until the people fill the vacancies by election as the legislature may direct.

Clause 3. This amendment shall not be so construed as to affect the election or term of any Senator chosen before it becomes valid as part of the Constitution.


12. The Federal Government does not collect Consumer Price Index or Cost of Living data in rural areas. The Bureau of Labor Statistics, in it's dataset 'Consumer Expenditure Survey', does track spending in rural vs. urban areas.

This chart is based upon annual earnings and net change of assets for the given year.

www.bls.gov/ces/data.htm cxubc and cxunx series data

13. Easy-Bake® is a registered trademark owned by Hasbro, Inc, Pawtucket, R.I.

14. http://www.nws.noaa.gov/climate/local_data.php?wfo=lwx

15. http://www.ncdc.noaa.gov/paleo/globalwarming/images/temperature-change.jpg

16. http://cdiac.ornl.gov/ftp/trends/temp/jonescru/global.dat

17. http://www.esrl.noaa.gov/gmd/ccgg/iadv/index.php

18. http://data.giss.nasa.gov/gistemp/graphs_v3/

19. "Past Climate Variability and Change in the Arctic and at High Latitudes," U.S. Climate Change Science Program Synthesis and Assessment Product 12 January 2009
Lead Agency U.S. Geological Survey Contributing Agencies National Oceanic and Atmospheric Administration, National Science Foundation

20. ftp://cdiac.ornl.gov/pub/trends/temp/vostok/vostok.1999.temp.dat
http://www.ncdc.noaa.gov/paleo/globalwarming/paleodata.html

21. Weather, Climate, and Paleoclimatology, NOAA.gov, Aug. 2008

22. ftp://ftp.ngdc.noaa.gov/STP/SOLAR_DATA/SOLAR_IRRADIANCE/ARVESEN.DAT

23. http://nssdc.gsfc.nasa.gov/planetary/factsheet/
(Stefan-Boltzman Law and Ideal Gas Law applied to data.)

24. Ibid.

25. http://pds.jpl.nasa.gov/planets/captions/venus/
sapasmon.htm

26. http://en.wikipedia.org/wiki/File:Phanerozoic_Sea
_Level.png

27. http://soundwaves.usgs.gov/2002/01/meetings2.html

28.
http://deltas.usgs.gov/presentations/Burkett,%20Virginia.pdf
(page 13 sea levels)

29.
http://en.wikipedia.org/wiki/File:Sea_level_temp_140ky.gif

30. Figure 1.5 "Long-Term Aridity Changes in the West",
U.S. Climate Change Science Program, Synthesis and
Assessment Report 3.4 "Abrupt Climate Change", U.S.
Geological Survey April 25, 2008
http://downloads.climatescience.gov/sap/sap3-4/sap3-4-
final-report-all.pdf

31. http://www.nas.nasa.gov/About/Education/Ozone/
history.html

32. http://www.ozonelayer.noaa.gov/science/basics.htm

www.ingramcontent.com/pod-product-compliance
Lightning Source LLC
Chambersburg PA
CBHW051502170526
45166CB00001B/354